Memoirs of the American Mathematical Society

Number 206

Richard A

Elements ... analysis

Published by the

AMERICAN MATHEMATICAL SOCIETY

Providence, Rhode Island

VOLUME 16 · ISSUE 1 · NUMBER 206 (second of 4 numbers) · SEPTEMBER 1978

[redacted]

AMS (MOS) 1970 subject classifications. Primary 58D15, 57A20, 58B10, 58C20, 46G05, 58E05, 49F15; Secondary 46E15, 46A99, 58B05.

Library of Congress Cataloging in Publication Data CIP

Graff, Richard A 1944-
Elements of non-linear functional analysis.

(Memoirs of the American Mathematical Society ; no. 206)
"Volume 16, issue 1 ... second of 4 numbers."
Based on the author's thesis.
Bibliography: p.
1. Functional analysis. I. Title. II. Series: American Mathematical Society. Memoirs ; no. 206.
QA3.A57 no. 206 [QA320] 510'.8s [515'.7]
ISBN 0-8218-2206-3 78-14727

ABSTRACT

A new notion of Frechet differentiability is introduced for maps between Banach spaces which are dual spaces, and it is shown that diffeomorphisms in the class of mappings thus isolated preserve the bounded weak-star topology as well as the ordinary metric topology. Basic Banach manifold differential topology is redeveloped under the assumption that the transition functions between the coordinate charts possess this refined type of differentiability, and it is pointed out that such manifolds possess additional structures not found on general manifolds: for example, a globally defined weaker topology, and Finsler metrics in which boundedness is equivalent to compact closure in the global weaker topology. It is shown that the usual manifolds of maps used in global analysis are manifolds of this type, and infinite-dimensional Morse theory and Lusternik-Schnirelman critical point theory are developed using these additional structures. An application is made to a general class of nonlinear problems in the calculus of variations. Also, an abstract inverse function theorem is presented for the class of differentiable mappings introduced in this paper. It states that, if the derivative of a mapping is an isomorphism at a given point, then the map is a local diffeomorphism on a neighborhood of the point, where the neighborhood and its image are open in the bounded weak-star topology.

TABLE OF CONTENTS

		Page
	INTRODUCTION	vi
1.	SECTION FUNCTORS	1
2.	THE BOUNDED WEAK* TOPOLOGY	22
3.	DIFFERENTIAL CALCULUS	42
4.	DIFFERENTIABLE FUNCTIONS ON bw* SPACES	71
5.	AN INVERSE FUNCTION THEOREM	81
6.	$\mathcal{U}$-MAPS AND THE GEOMETRY OF FUNCTION SPACES	103
7.	INFINITE-DIMENSIONAL DIFFERENTIABLE MANIFOLDS	127
8.	CRITICAL POINT THEORY ON Bw* MANIFOLDS	147
9.	AN APPLICATION TO THE CALCULUS OF VARIATIONS	181
10.	BIBLIOGRAPHY	194

INTRODUCTION.

For the past two decades it has been known that many of the function spaces which arise naturally in the investigation of problems involving non-linear partial differential operators possess natural Banach manifold differentiable structures, and that C^{∞} non-linear differential operators often induce differentiable maps between appropriate function space manifolds. In this way, the tools of differential calculus and infinite-dimensional differential topology can be brought to bear on such problems. However, such tools by themselves are rarely adequate, and in situations where infinite-dimensional differential topology can be successfully applied it is usually necessary to use additional properties which function spaces and operator-induced morphisms possess, but which have no abstract equivalents in Banach manifold differential topology. Indeed, some global analysts insist that it is useless to consider an abstract model at all, that global analysts should forget about abstractions and work exclusively with function spaces and differential operators.

However, the purpose of this monograph is to take a third approach: while conceding that Banach manifold differential topology is too weak an abstract setting for global analysis, I present herein a more highly structured category whose objects seem to possess many of the properties relevant to concrete applications. Many function spaces will be seen to be examples of these new abstract objects; and as an illustration of the suitability of this model, several abstract theorems will be presented which imply the basic results of Palais and Smale in non-linear variational calculus.

The key to this theory is the bw* space. The simplest (though not most elegant) way to describe a bw* space is as the dual space of a Banach space, endowed not with its metric topology, but rather with the strongest topology which agrees with the weak-star topology on bounded subsets. An important subclass, the Bw* spaces, consists of dual spaces of separable Banach spaces (Bw* spaces differ topologically from other bw* spaces in significant ways, cf. Chapter 2). These bw* spaces are locally convex, and the category of C^{∞} Bw* manifolds and differentiable maps is what I propose as the new and improved abstract model to replace Banach manifold theory.

Since infinite-dimensional bw* manifolds are not metrizable, an immediate question to ask is: in what sense is it possible to refer to a "C^{∞} map" between two such spaces? An attempt to lend meaning to this concept led to the material in Chapter 3. In Chapter 3, I show that the most elementary techniques of Banach space differential calculus can be extended to a class of locally convex spaces slightly larger than the class of Banach spaces. For a space to be in this class, it need not be metrizable: but spaces of linear and multi-linear maps from it to Banach spaces must have certain properties similar to properties of multi-linear maps between Banach spaces. This enlarged class of locally convex spaces is not large, but it does include all normed spaces, certain Frechet spaces, and most importantly - bw* spaces. The basic result of Chapter 3 is that, for each $k \in N$, the class of C^k maps between such linear spaces is closed under composition.

With this theory of differential calculus it is possible to investigate differentiable maps between bw* spaces. It is shown in Chapter 4 that, for each $k \in N$, a C^k map between two bw* spaces is automatically C^{k-1} in the usual sense of Banach space differential calculus if the bw* spaces under consideration are re-equipped with their metric linear structures. Furthermore, since bounded subsets of a bw* space have compact closure, a C^k map between bw* spaces is bounded on bounded subsets of its domain, and similarly each continuous derivative of the map is bounded on bounded sets.

Thus the class of C^∞ maps between two bw^* spaces actually represents a subclass of the class of C^∞ maps between the associated Banach spaces. However, in order to isolate this particular subclass, it was necessary to take a roundabout route through a generalized differential calculus for locally convex spaces.

Assume for the moment that M is a C^∞ Bw^* manifold modeled on a Bw^* space X. Then each chart (U,φ) on M gives a homeomorphism of an open subset U in M with an open subset $\varphi(U)$ in X. Since $\varphi(U)$ is open in X, it will still be open if we re-equip X with its Banach space topology. And if we let (V,ψ) be another chart on M, then $\psi\circ\varphi^{-1} : \varphi(U \cap V) \to \psi(U \cap V)$ is a C^∞ diffeomorphism between open subsets of a Bw^* space, and hence will still be a C^∞ diffeomorphism between open subsets if X is re-equipped with its metric linear structure. Thus the atlas on M can be used as the atlas for the differentiable structure of a C^∞ Banach manifold on a retopologized version of M, i.e. every C^∞ Bw^* manifold automatically carries in addition a secondary structure as a Banach manifold. It is shown in Chapter 6 that spaces of Sobolev maps, Hölder maps, and Lipschitz maps are examples of C^∞ Bw^* manifolds, and so one might easily conclude that in the past global analysts have concerned themselves with a secondary differentiable structure on function space manifolds while remaining ignorant of the primary one.

In Chapter 7, the rudiments of Bw^* manifold theory are presented. It is shown that the natural map between a Bw^* manifold and its associated Banach manifold is a homotopy equivalence. A natural class of Finsler structures on the associated Banach manifold is introduced and it is shown that, with respect to the metrics which arise from such Finsler structures, the sets which are bounded in metric are precisely those which have compact closure in the Bw* topology.

In Chapter 8, versions of Morse Theory and Lusternik-Schnirelman critical point theory tailored for Bw* manifolds are developed, and several abstract theorems concerned with critical points of non-linear functionals are proved. And in the last chapter, these theorems are applied to function spaces and non-linear Lagrangians to obtain those theorems in the calculus of variations which have come to be known as Palais-Smale theory.

About the first two chapters: Chapter 2 contains a basic exposition of the facts concerning bw* spaces which are used elsewhere in this paper, Chapter 1 consists basically of motivational material. In Chapter 1, it is shown that many function spaces are topological Bw* manifolds, and an indication is given of why it is natural to expect these Bw* manifolds to be differentiable (and hence, why it was natural to expect the existence of a theory of differential calculus for bw* spaces). Also, the reader should be aware that techniques developed in Chapter 1 are used again and again in Chapters 2, 4, and 5, and that in Chapter 1 it may be seen where the ideas for these techniques originated.

A natural question to ask is whether all the basic theorems of Banach space differential calculus have analogues in bw* differential calculus. For instance, if X is a bw* space, and $f : X \to X$ is a C^∞ map for which $Df(0)$ is an invertible linear map, is f locally invertible? Note that, since f is automatically C^∞ in the usual Banach space sense, the usual inverse function theorem implies that f is locally invertible on a norm-open neighborhood of zero. However, it is much more to ask whether or not f is an isomorphism on a bw*-neighborhood of zero, since bw*-open sets are unbounded. Similarly, a C^∞ vector field on X is automatically Lipschitz in the metric linear structure and hence generates a flow in the usual sense; however, it is quite another thing to ask whether the domain of this flow is open in $\mathbb{R} \times X$ in the bw* topology, and if so, whether the flow is weakly continuous or differentiable.

The answer to each of the above questions is: No. However, in addition to introducing C^k maps between bw* spaces in this paper, I introduce another class of maps between bw* spaces in Chapters 5 and 6 whose derivatives, besides being continuous in the bw* sense, are assumed to satisfy a condition which reduces to the usual local Lipschitz condition if the domain and range are finite-dimensional. A map with k such derivatives is referred to as a $\mathcal{U}^k$ map after Karen Uhlenbeck, who first noticed that a slight variant of this condition is satisfied by the second derivative of any change of coordinate charts on Sobolev manifolds. These $\mathcal{U}^k$ maps are essential to the applications developed in Chapters 8 and 9 (indeed, Sobolev manifolds are shown in Chapter 6 to be, not merely C^∞ Bw* manifolds, but also $\mathcal{U}^\infty$ Bw* manifolds), and the class of $\mathcal{U}^k$ maps is closed under composition. So it is natural to ask the questions about the inverse function theorem and the existence of flows for vector fields once more, this time for $\mathcal{U}^k$ maps. And in this revised context the answer to each question is: yes. A $\mathcal{U}^k$ map $f : X \to X$ whose derivative at zero is invertible does have a local $\mathcal{U}^k$ inverse on a neighborhood of zero, and a $\mathcal{U}^k$ vector field does generate a $\mathcal{U}^k$ flow. In addition, if U is any open subset of a Bw* space X (and here is a significant difference between Bw* spaces and other bw* spaces), then there is a $\mathcal{U}^\infty$ real-valued function $\alpha : X \to R$ such that $U = \{x \in X : \alpha(x) \neq 0\}$, which implies the strongest possible result about $\mathcal{U}^\infty$ partitions of unity on paracompact $\mathcal{U}^\infty$ Bw* manifolds.

The inverse function theorem for $\mathcal{U}^k$ maps is proven in Chapters 5 and 6, and the existence of $\mathcal{U}^\infty$ partitions of unity in Chapters 4 and 6.

As for the proof that $\mathcal{U}^k$ vector fields generate $\mathcal{U}^k$ flows, it will appear in a later article together with an application which will show that any reasonable space of maps from a smooth compact manifold M to a $\mathcal{U}^\infty$ Bw* manifold N is again a Bw* manifold.

The first six chapters of this paper consist of an updated version of my Ph.D. dissertation. Chapter 7 and 8 are slightly revised versions of material which was developed in 1971. Indeed, the purpose of the original, somewhat primitive version of the analysis in Chapters 3 through 5 was to prove Theorem 8.57 (part of the conjecture of 8.57 involved the assumption that the coordinate transformations on Sobolev manifolds are Bw*-differentiable, together with the assumption that Bw* differentiability is related to Banach differentiability as described in Chapter 4). By the time I began the actual writing of the dissertation, however, the development of differential calculus in Bw* spaces had progressed to the point where it seemed that this material might be more significant than the applications to the calculus of variations. So I wrote my dissertation with the idea in mind of later using it as the first six chapters of a basic textbook on Bw* differential topology and applications to global analysis.

Having learned since then that there is no demand for such a text, and in the interest of keeping this monograph as short as possible, I have omitted a few topics which I had originally intended to include. Thus the reader will find in Chapter 1 that I promise an appendix which will show how to "complete" manifolds of maps which come from certain non-compact section functors to obtain Bw* manifolds. However, this appendix has been omitted; it may appear later as a separate publication.

Similarly, at several points in the text I promise examples of Bw* manifolds of maps which are C^{∞} but not $\mathcal{U}^{\infty}$. This class of examples has also been omitted, and will appear later in the paper that $\mathcal{U}^{k}$ vector fields generate $\mathcal{U}^{k}$ flows.

A word about the origin of the idea that Banach manifolds of maps can have a globally-defined weaker topology: in the summer of 1969, Karen Uhlenbeck and James Dowling verified independently that weakly convergent sequences are invariant under a change of naturally-arising coordinate charts on a Sobolev manifold. Their methods involved inequalities relating norms of distributional derivatives, and were somewhat more complicated than the topological approach presented here in Theorem 1.25. The ω-construction of Theorem 1.25, which associates a globally-defined weaker topology to a Banach manifold of maps under much more general conditions than those original methods, was something I noticed a week after the initial discovery by the above-mentioned researchers. In fact, it was an attempt to find an intuitive justification for the existence of the topology discovered by Uhlenbeck and Dowling which led me to the ω-construction and, later, to the material in Chapter 2.

I would like to acknowledge the contributions of several individuals who provided valuable information at several stages in the development of this material. R. R. Phelps and J. H. C. Whitfield provided me with capsule summaries of the basic facts about differentiable norms on Banach spaces which I needed to extend the critical point theory of Chapter 8 from Hilbert manifolds to more general Banach manifolds. J. P. Penot developed the proof of Lemma 5.25 presented here, which was necessary for the extension of the inverse function theorem of Chapter 5 from Bw* spaces to general bw* spaces (my original proof of the inverse function theorem for Bw* spaces only required a proof of Lemma 5.25 for metric spaces). J. R. Dorroh helped me simplify and extend the proof of Lemma 5.14, which I had originally proved for bw* spaces, so that it now applies to all locally convex spaces. And, most significantly of all, R. S. Palais provided me with suggestions, insights about global analysis, and general encouragement during the period 1969-1972, when I was a graduate student and much of the development of this theory took place.

ELEMENTS OF NON-LINEAR FUNCTIONAL ANALYSIS

by R. A. Graff

1. SECTION FUNCTORS

Let M be a smooth compact manifold, possibly with boundary.

<u>1.1 Definition.</u> $FB(M)$ is the category whose objects are smooth finite-dimensional fiber bundles over M (the fibers are manifolds without boundaries) and whose morphisms are defined as follows: if E_1, E_2 are objects of $FB(M)$, then

$$\mathrm{Map}(E_1,E_2) = \{C^\infty \text{ fiber-preserving maps from } E_1 \text{ to } E_2\} .$$

<u>1.2 Definition</u>. A section functor $\mathcal{M}$ on $FB(M)$ is a functor from $FB(M)$ to the category of topological spaces and continuous maps such that:

(1) For each bundle E over M, the points of $\mathcal{M}(E)$ are continuous sections of E, and the natural inclusion $\mathcal{M}(E) \to C^0(E)$ is continuous.

(2) For bundles E_1, E_2 over M, and a morphism $f \in \mathrm{Map}(E_1,E_2)$, $\mathcal{M}(f)(s) = f\circ s$ for all $s \in \mathcal{M}(E_1)$.

Note that we have (by definition) a natural transformation $\mathcal{M} \longrightarrow C^0$.

Our first goal is to show that the local structure of $\mathcal{M}(E)$ depends only on the local structure of E. In order to do this, it will be necessary to introduce a category which is related to $FB(M)$.

<u>1.3 Definition</u>. $FVB(M)$ is the category whose objects are smooth vector bundles over M, and whose morphisms are defined as follows: if ξ, η are smooth vector bundles over M, then

$$\mathrm{Map}(\xi,\eta) = \{C^\infty \text{ fiber-preserving maps from } \xi \text{ to } \eta\}.$$

Note that $FVB(M)$ is a full subcategory of $FB(M)$.

Received by the editor June 29, 1977.

1.4 Definition. A section functor on FVB(M) is a functor $\mathcal{M}$ from FVB(M) to the category of topological spaces and continuous maps such that:

(1) For each bundle ξ over M, the points of $\mathcal{M}(\xi)$ are continuous sections of ξ, and the natural inclusion $\mathcal{M}(\xi) \to C^0(\xi)$ is continuous.

(2) For bundles ξ, η over M, and $f \in \mathrm{Map}(\xi,\eta)$, $\mathcal{M}(f)(s) = f \circ s$ for all $s \in \mathcal{M}(\xi)$.

1.5 Lemma. Let $\mathcal{M}$ be a section functor on either FB(M) or FVB(M), and let ξ be a bundle over M. If U is an open subset of ξ, then $\{s \in \mathcal{M}(\xi) \mid s(M) \subset U\}$ is an open subset of $\mathcal{M}(\xi)$.

Proof. Since the topology on $C^0(\xi)$ is the compact-open topology, $\{s \in C^0(\xi) \mid s(M) \subset U\}$ is an open subset of $C^0(\xi)$. Since by assumption the map $\mathcal{M}(\xi) \to C^0(\xi)$ is continuous, $\{s \in \mathcal{M}(\xi) \mid s(M) \subset U\}$ is open in $\mathcal{M}(\xi)$. □

To demonstrate the desired structure theorem, it will be necessary to have the concept of open vector subbundles of a fiber bundle, and a lemma about their existence.

1.6 Definition. Let E be a fiber bundle over M. An open vector subbundle ξ of E is a smooth vector bundle whose underlying set is open in E, such that the natural inclusion $\xi \longrightarrow E$ is a bundle morphism.

Note that an open subset of E which is the underlying manifold of an open vector subbundle, will be the underlying manifold of many distinct open vector subbundles (all canonically diffeomorphic, but with distinct linear structures). We will abuse the definition of vector bundle slightly and say that a vector bundle ξ is contained in a fiber bundle E if the underlying set of ξ is a subset of E, and if the natural inclusion $\xi \longrightarrow E$ is a bundle morphism.

1.7 Lemma. Let $s \in C^0(E)$, and let U be an open subset of E such that $s(M) \subset U$. Then there exists an open vector subbundle

ξ of E such that $\xi \subset U$ and $s(M) \subset \xi$.

Proof. This is finite-dimensional differential topology, and should be familiar to those readers who are acquainted with sprays. The original derivation, in Chapter 12 of [33], is clear, concise and self-contained, so we will not present a proof here. □

1.8 Lemma. Let $\mathcal{M}$ be a section functor (on either FB(M) or FVB(M)), ξ and η bundles of dimension q over M, and $f : \xi \to \eta$ a smooth embedding. Then $\mathcal{M}(f)$ is a continuous embedding of $\mathcal{M}(\xi)$ onto an open subset of $\mathcal{M}(\eta)$.

Proof. Let $s_o \in \mathcal{M}(\eta)$ such that $s_o(M) \subset f(\xi)$. Then $f^{-1} \circ s_o \in C^o(\xi)$. If ξ is not a vector bundle, choose an open vector subbundle $\tilde{\xi}$ of ξ such that $f^{-1} \circ s_o(M) \subset \tilde{\xi}$. Note that $f(\tilde{\xi})$ is an open vector subbundle of η.

Since $s_o(M)$ is a compact subset of $f(\tilde{\xi})$, and since $f(\tilde{\xi})$ is open in η, there exists a C^∞ function α defined on η such that $\alpha = 1$ on an open neighborhood V of $s_o(M)$, and such that the support of α is contained in $f(\tilde{\xi})$. Define $g : \eta \to \xi$ by

$$g(y) = \begin{cases} \alpha(y) f^{-1}(y), & \text{if } y \in f(\tilde{\xi}) \\ 0_{p(y)}, & \text{if } y \notin f(\tilde{\xi}) \end{cases}$$

where $p : \tilde{\xi} \to M$ is the bundle projection.

Then $\mathcal{M}(g)(s_o) = f^{-1} \circ s_o \in \mathcal{M}(\xi)$, so that the image of $\mathcal{M}(f)$ equals $\{s \in \mathcal{M}(\eta) \mid s(M) \subset f(\xi)\}$, and is thus an open subset of $\mathcal{M}(\eta)$ by Lemma 1.5. Further, $\{s \in \mathcal{M}(\eta) \mid s(M) \subset V\}$ is an open subset of $\mathcal{M}(\eta)$ which contains s_o. Since $\mathcal{M}(g)$ restricted to this set is the inverse of $\mathcal{M}(f)$ restricted to $\{s \in \mathcal{M}(\xi) \mid s(M) \subset f^{-1}(V)\}$, $\mathcal{M}(f)$ is locally a homeomorphism with its image. But $\mathcal{M}(f)$ is obviously an injection, so that $\mathcal{M}(f)$ is a homeomorphism onto its image. □

It is now possible to prove the structure theorem for section functors on FB(M) mentioned after Definition 1.2. Note that we have a natural functor S from FVB(M) to FB(M). For each vector bundle ξ over M, $S(\xi)$ is the underlying manifold of ξ. If ξ and η

are vector bundles, and $f : \xi \to \eta$, $S(f) = f$. Thus each section functor $\mathcal{M}$ on FB(M) induces a section functor $\mathcal{M}\circ S$ on FVB(M) by restriction.

1.9 Theorem. The restriction mapping from functors on FB(M) to functors on FVB(M) is a 1-1 correspondence.

Proof. To show that the mapping is surjective, let $\mathcal{M}$ be a section functor on FVB(M). We will define a functor $\mathfrak{F}$ on FB(M) such that $\mathfrak{F}\circ S = \mathcal{M}$.

Assume E is a fiber bundle over M, and let $A = \{\xi \mid \xi$ is an open vector subbundle of $E\}$. Set $\mathfrak{F}(E) = \bigcup_{\xi\in A} \mathcal{M}(\xi)$, with the strongest topology such that $i_\xi : \mathcal{M}(\xi) \to \mathfrak{F}(E)$ is continuous for all $\xi \in A$. Note that all maps $\mathcal{M}(\xi) \to C^0(E)$ are continuous, so the map $\mathfrak{F}(E) \to C^0(E)$ is continuous.

To finish the proof, it suffices to show that $\mathcal{M}(\xi)$ is an open subset of $\mathfrak{F}(E)$ for each $\xi \in A$, and that the topology induced on $\mathcal{M}(\xi)$ as a subset of $\mathfrak{F}(E)$ coincides with the original topology on $\mathcal{M}(\xi)$. For, assuming this to be true, let E_1, E_2 be fiber bundles over M, $f \in \mathrm{Map}(E_1, E_2)$, and $s \in \mathfrak{F}(E_1)$. We will show that $\mathfrak{F}(f)$ is continuous by showing there exists a neighborhood U of s in $\mathfrak{F}(E_1)$ such that $\mathfrak{F}(F)|_U$ is continuous. Choose an open vector subbundle ξ of E_1 such that $s \in \mathcal{M}(\xi)$. Since $f\circ s \in C^0(E_2)$, there exists an open vector subbundle η of E_2 such that $f\circ s(M) \subset \eta$. Since $f^{-1}(\eta)$ is open in E_1, there exists an open vector subbundle $\tilde{\xi}$ of E_1 such that $s(M) \subset \tilde{\xi} \subset \xi \cap f^{-1}(\eta)$. Since $\tilde{\xi}$ is also an open vector subbundle of ξ, $s \in \mathcal{M}(\tilde{\xi})$ by Lemma 1.8. But $\mathcal{M}(\tilde{\xi})$ is an open subset of $\mathfrak{F}(E_1)$, and the map $\mathfrak{F}(f)|_{\mathcal{M}(\tilde{\xi})} = \mathcal{M}(f) : \mathcal{M}(\tilde{\xi}) \to \mathcal{M}(\eta) \subset \mathfrak{F}(E_2)$ is continuous, $\Longrightarrow \mathfrak{F}(f)$ is continuous. And obviously $\mathfrak{F}\circ S = \mathcal{M}$

To see that the subspace topology on $\mathcal{M}(\xi)$ has the desired properties, let U be an open subset of $\mathcal{M}(\xi)$ in the original topology. To show that U is open in $\mathfrak{F}(E)$, it is sufficient to show that $U \cap \mathcal{M}(\eta)$ is open in $\mathcal{M}(\eta)$ for any other open vector subbundle η of E. So let $s \in U \cap \mathcal{M}(\eta)$. Then $s(M) \subset \xi \cap \eta$, ==> there exists an open vector subbundle γ of E such that $s(M) \subset \gamma \subset \xi \cap \eta$. Now the inclusion $\mathcal{M}(\gamma) \to \mathcal{M}(\xi)$ is a continuous embedding by Lemma 1.8, ==> $U \cap \mathcal{M}(\gamma)$ is open in $\mathcal{M}(\gamma)$. But $\mathcal{M}(\gamma)$ is also embedded as an open subspace of $\mathcal{M}(\eta)$, so that $U \cap \mathcal{M}(\gamma)$ is an open neighborhood of s in $\mathcal{M}(\eta)$. Thus $U \cap \mathcal{M}(\eta)$ is open in $\mathcal{M}(\eta)$ ==> U is open in $\mathfrak{F}(E)$, ==> $\mathcal{M}(\xi)$ is an open subset of $\mathfrak{F}(E)$, and the subspace topology on $\mathcal{M}(\xi)$ coincides with the original topology.

To show that the mapping of functors is injective, it suffices to notice that, if $\mathcal{M}$ and $\mathcal{N}$ are two distinct functors on FB(M), then there exists a fiber bundle E such that $\mathcal{M}(E) \neq \mathcal{N}(E)$. But by Lemmas 1.7 and 1.8, $\mathcal{M}(E) = \bigcup_{\xi \in A} \mathcal{M}(\xi)$, $\mathcal{N}(E) = \bigcup_{\xi \in A} \mathcal{N}(\xi)$, where $A = \{$open vector subbundles of E$\}$.

If $\mathcal{M}(\xi) = \mathcal{N}(\xi)$ for all $\xi \in A$, $\mathcal{M}(E) = \mathcal{N}(E)$. Thus there exists $\xi_o \in A$ such that $\mathcal{M}(\xi_o) \neq \mathcal{N}(\xi_o)$, ==> $\mathcal{M} \circ S \neq \mathcal{N} \circ S$. □

The above theorem makes it possible to refer to a section functor "on M", and to regard it as defined on FB(M) or FVB(M) according to our needs of the moment.

It is natural to ask how many sections must be in the spaces $\mathcal{M}(E)$. Since the functor which assigns the empty set to each bundle E over M satisfies the definition of a section functor, we see: not many. However, if $\mathcal{M}$ is any other section functor on FB(M), then it is easy to see that $C^\infty(E) \subset \mathcal{M}(E)$ for every bundle E over M. To see that this is true, let E_o be a bundle such that $\mathcal{M}(E_o) \neq \emptyset$, and let $s_o \in \mathcal{M}(E_o)$. Then, for each bundle E over M and $s \in C^\infty(E)$, define $f_s \in \mathrm{Map}(E_o, E)$ by $f_s = s \circ p_o$, where $p_o : E_o \to M$ is the bundle projection map. Note that, since f_s is the composition of C^∞ maps, it is C^∞. So $\mathcal{M}(f_s)$ is a map from $\mathcal{M}(E_o)$ to $\mathcal{M}(E)$. But $\mathcal{M}(f_s)(s_o) = f_s \circ s_o = s \circ p_o \circ s_o = s \circ \mathrm{Id}_M = s$, and so we see that $s \in \mathcal{M}(E)$.

Thus we have a natural transformation from C^∞ to $\mathcal{M}$ for every section functor except the trivial one. It is natural to conjecture that this natural transformation is continuous (i.e. that we have continuous inclusions $C^\infty(E) \to \mathfrak{M}(E) \to C^\circ(E)$ for every bundle E) . However, this is merely conjecture at the moment.

From now on, we will assume without further comment that every section functor is non-trivial (and hence, that $C^\infty(E) \subset \mathfrak{M}(E)$ for every bundle E) .

We have now come as far as we can without assuming additional structure on the objects $\mathfrak{M}(E)$. Theorem 1.9 suggests that, if we wish to strengthen the local structure on the objects $\mathfrak{M}(E)$, it is sufficient to assume additional structure on the objects $\mathfrak{M}(\xi)$, ξ a vector bundle over M. If we notice that the continuous sections of a vector bundle form a vector space, it becomes natural to consider section functors $\mathfrak{M}$ such that $\mathfrak{M}(\xi)$ is a vector space for each vector bundle ξ .

<u>1.10 Definition.</u> A LTS (linear topological space) section functor $\mathfrak{M}$ on M is a section functor on $FVB(M)$ such that, for each vector bundle ξ on M, $\mathfrak{M}(\xi)$ is a vector subspace of $C^0(\xi)$, and such that $\mathfrak{M}(\xi)$ is a linear topological space with respect to its topology and this vector space structure.

When we refer to a product in this paper, we will always mean the product in the appropriate category. For example, in the following proposition, $E_1 \times E_2$ means the fibered product of E_1 and E_2 over M, i.e. the product of E_1 and E_2 in the category $FB(M)$.

<u>1.11 Proposition</u>. Let $\mathfrak{M}$ be a LTS section functor on M. Then for any two fiber bundles E_1, E_2 over M, $\mathfrak{M}(E_1 \times E_2) = \mathfrak{M}(E_1) \times \mathfrak{M}(E_2)$.

<u>Proof</u>. We have smooth projections $p_i : E_1 \times E_2 \to E_i$, so we have maps $\mathfrak{M}(p_i) : \mathfrak{M}(E_1 \times E_2) \to \mathfrak{M}(E_i)$, and hence a map $\mathfrak{M}(E_1 \times E_2) \to \mathfrak{M}(E_1) \times \mathfrak{M}(E_2)$. This map exists for any section functor, and is obviously injective. We will use our additional local structure to show that the map is surjective and has a continuous inverse. So assume $s_i \in \mathfrak{M}(E_i)$, and let ξ_i be an open vector bundle neigh-

borhood of s_i in E_i. Then $\xi_1 \times \xi_2$ is an open vector subbundle of $E_1 \times E_2$. We have a natural smooth map $\begin{matrix}\xi_1 \to \xi_1 \times \xi_2\\ v \to (v,0)\end{matrix}$, and a natural map $\begin{matrix}\xi_2 \to \xi_1 \times \xi_2\\ w \to (0,w)\end{matrix}$. These induce maps $\begin{matrix}\mathcal{M}(\xi_1) \to \mathcal{M}(\xi_1 \times \xi_2)\\ t_1 \to (t_1,0)\end{matrix}$ and $\begin{matrix}\mathcal{M}(\xi_2) \to \mathcal{M}(\xi_1 \times \xi_2)\\ t_2 \to (0,t_2)\end{matrix}$. Since addition is a continuous operation in $\mathcal{M}(\xi_1 \times \xi_2)$, the map $\begin{matrix}\mathcal{M}(\xi_1) \times \mathcal{M}(\xi_2) \to \mathcal{M}(\xi_1 \times \xi_2)\\ t_1 \times t_2 \to (t_1,t_2)\end{matrix}$ is continuous. Note also that this implies that $(s_1,s_2) \in \mathcal{M}(E_1 \times E_2)$. Since the map $\mathcal{M}(E_1 \times E_2) \to \mathcal{M}(E_1) \times \mathcal{M}(E_2)$ is surjective, and has a local inverse, it is a homeomorphism. □

<u>1.12 Remark</u>. Let $\mathcal{M}$ be a LTS section functor on M. Then $\mathcal{M}(\mathbb{R})$ (i.e. $\mathcal{M}(M \times \mathbb{R})$) is a topological algebra under pointwise multiplication, since for $\begin{matrix} m : \mathbb{R} \times \mathbb{R} \to \mathbb{R}\\ a,b \to ab\end{matrix}$

$\mathcal{M}(\mathbb{R}) \times \mathcal{M}(\mathbb{R}) \simeq \mathcal{M}(\mathbb{R} \times \mathbb{R}) \xrightarrow{\mathcal{M}(m)} \mathcal{M}(\mathbb{R})$.

More generally, let ξ and η be two vector bundles over M, and let r be a positive integer. Then the natural map

$$\mathcal{M}(L^r(\xi,\eta)) \times \underbrace{\mathcal{M}(\xi) \times \ldots \times \mathcal{M}(\xi)}_{r \text{ times}} \to \mathcal{M}(\eta)$$

is continuous. Thus each element T of $\mathcal{M}(L^r(\xi,\eta))$ induces an element $\tilde{T}$ of $L^r(\mathcal{M}(\xi),\mathcal{M}(\eta))$, defined as follows:

$$(\tilde{T}(s_1,\ldots,s_r))(x) = T(x)(s_1(x),\ldots,s_r(x)) \quad \text{for all } x \in M.$$

Assume now that $\mathcal{M}(L^r(\xi,\eta))$, $\mathcal{M}(\xi)$, and $\mathcal{M}(\eta)$ are locally convex spaces. Then the continuity of the evaluation map implies that, given a semi-norm ν on $\mathcal{M}(\eta)$, there exists semi-norms μ on $\mathcal{M}(\xi)$, λ on $\mathcal{M}(L^r(\xi,\eta))$ such that $\nu(T(s_1,\ldots,s_r)) \leq \lambda(T)\cdot\mu(s_1)\cdot\ldots\cdot\mu(s_r)$. Now the space $L^r(\mathcal{M}(\xi),\mathcal{M}(\eta))$ also has a natural topology: the topology of uniform convergence on bounded subsets of $\mathcal{M}(\xi) \times \ldots \times \mathcal{M}(\xi)$. Thus the continuity of the evaluation map implies that the injection $\mathcal{M}(L^r(\xi,\eta)) \to L^r(\mathcal{M}(\xi),\mathcal{M}(\eta))$ is continuous.

The above remark will be of use later when we discuss the differentiability of fiber-induced maps. For the moment, we return to the classification of section functors. Let n be a fixed positive integer.

1.13 Definition. FVB(n) is the category whose objects are smooth finite-dimensional vector bundles over compact n-dimensional manifolds (possibly with boundary), and whose morphisms are defined as follows:

Let ξ be a vector bundle over M, η a vector bundle over N.

(1) If $M \neq N$, $\mathrm{Map}(\xi,\eta) = \emptyset$

(2) If $M = N$, $\mathrm{Map}(\xi,\eta) = \{C^\infty$ fiber-preserving maps from ξ to $\eta\}$.

1.14 Definition. FB(n) is defined as above, replacing "vector bundles" with "fiber bundles".

1.15 Definition. A section functor on FVB(n) (resp. FB(n)) is a functor from FVB(n) (resp. FB(n)) to the category of topological spaces and continuous maps such that the restriction of the functor to FVB(M) (resp. FB(M)) is a section functor on M for each compact n-dimensional manifold M.

1.16 Remark. Obviously section functors on FVB(n) and FB(n) are in natural 1-1 correspondence, so that we may simply refer to "n-dimensional section functors".

Our next goal will be to find a fairly general class of n-dimensional section functors $\mathcal{M}$ which are classified by $\mathcal{M}(D^n,\mathbb{R})$, where D^n is the n-dimensional disc, and to find conditions under which we may take a locally convex space $\mathcal{N}(D^n,\mathbb{R})$ of continuous functions on the unit disc and extend it to an n-dimensional section functor. This was actually done by Palais in [33] for n-dimensional Banach section functors, and his proofs, which we present here without change, will work in general. But first some definitions:

1.17 Definition. A category $\mathcal{C}$ of locally convex linear topological spaces will be called a CG category if:

(1) For each pair of spaces X and Y in $\mathcal{C}$, $X \times Y$ is in $\mathcal{C}$.

(2) If X is in $\mathcal{C}$ and Z is a closed subspace of X, then Z is in $\mathcal{C}$.

(3) If X and Y are in $\mathcal{C}$, and $f \in \mathrm{Map}(X,Y)$ is a bijection, then f is an isomorphism (i.e. the closed graph theorem holds for morphisms of $\mathcal{C}$).

The standard example of a CG category is of course the category of Banachable spaces (i.e. the underlying linear topological spaces of Banach spaces - normable, but with no distinguished norm).

1.18 Definition. Let $\mathcal{C}$ be a CG category. An n-dimensional $\mathcal{C}$ section functor $\mathfrak{M}$ is a section functor on FVB(n) such that:

(1) $\mathfrak{M}(\xi)$ is an object of $\mathcal{C}$ for each object ξ of FVB(n).

(2) If M and N are compact n-dimensional manifolds, ξ a vector bundle over M, and $i : N \to M$ a smooth embedding, then $s|_N \in \mathfrak{M}(\xi|_N)$ for each $s \in \mathfrak{M}(\xi)$, and the linear map $\mathfrak{M}(\xi) \to \mathfrak{M}(\xi|_N)$, $s \to s|_N$ is a continuous surjection.

Motivation for Condition (2) may be found in the Whitney and Calderon extension theorems.

1.19 Lemma. Let M be a compact n-dimensional manifold, and let $N_1, \ldots, N_r$ be compact n-dimensional submanifolds whose interiors cover M. Assume $\mathfrak{M}$ is an n-dimensional $\mathcal{C}$ section functor, and ξ a vector bundle over M. Define

$$\widetilde{\mathfrak{M}}(\xi) = \{ (s_1, \ldots, s_r) \in \bigoplus_{i=1}^{r} \mathfrak{M}(\xi|_{N_i}) : s_i|_{N_i \cap N_j} = s_j|_{N_i \cap N_j} \},$$

and let $\Phi : \mathcal{M}(\xi) \to \widetilde{\mathcal{M}}(\xi)$ $\quad$ $s \to (s|_{N_1}, \ldots, s|_{N_r})$. Then Φ is an isomorphism of locally convex spaces.

<u>Proof</u>. We first need to know that $\widetilde{\mathcal{M}}(\xi)$ is a $\mathcal{C}$-space. To see this, note that we have a continuous map

$$\bigoplus_{i=1}^{r} \mathcal{M}(\xi|_{N_i}) \to \bigoplus_{i=1}^{r} C^{o}(\xi|_{N_i}).$$

Since $\widetilde{C}^{o}(\xi)$ is a closed subspace of $\bigoplus_{i=1}^{r} C^{o}(\xi|_{N_i})$, $\widetilde{\mathcal{M}}(\xi)$ is closed in $\bigoplus_{i=1}^{r} \mathcal{M}(\xi|_{N_i})$, and hence $\widetilde{\mathcal{M}}(\xi)$ is a $\mathcal{C}$-space.

The map Φ is obviously continuous, and is injective since the interiors of the N_i cover M. Thus to show that Φ is an isomorphism, it suffices to show that Φ is surjective. So let $(s_1, \ldots, s_r) \in \widetilde{\mathcal{M}}(\xi)$. For each i, choose a section t_i of ξ such that $t_i|_{N_i} = s_i$. Now since the interiors of the N_i cover M, we may choose a smooth partition of unity $\{\rho_i\}$, $1 \le i \le r$ such that the support of ρ_i is contained in N_i. For each i, the map $\overline{\rho}_i : \xi \to \xi$ $\quad (x,v) \to (x, \rho_i(x)\cdot v)$ is a smooth vector bundle morphism. Thus

$\mathcal{M}(\overline{\rho}_i)(t_i) = \rho_i t_i \in \mathcal{M}(\xi)$ for each i, $\Longrightarrow \Sigma\rho_i t_i \in \mathcal{M}(\xi)$. But obviously $\Phi(\Sigma\rho_i t_i) = (s_1, \ldots, s_r)$. $\square$

<u>1.20 Theorem.</u> Let $\mathcal{M}$ be an n-dimensional $\mathcal{C}$ section functor. Then $\mathcal{M}$ is completely determined by $\mathcal{M}(D^n, \mathbb{R})$. Conversely, let $\mathcal{C}$ be a CG category, and suppose that V is a locally convex space of continuous functions on D^n such that V belongs to $\mathcal{C}$ and such that:

(1) the inclusion $V \to C^{o}(D^n, \mathbb{R})$ is continuous.

(2) If $\varphi : D^n \to D^n$ is a smooth embedding, the induced map $V \to C^{o}(D^n, \mathbb{R})$ $\quad f \to f \circ \varphi$ factors through a continuous surjection $V \to V$.

(3) If $f : D^n \times \mathbf{R}^q \to D^n \times \mathbb{R}$ is a smooth fiber-preserving map over the disc, then the induced map $V \times \ldots \times V \to V$ is continuous.

Then there exists a (necessarily unique) n-dimensional $\mathcal{C}$ section functor $\mathcal{N}$ such that $\mathcal{N}(D^n,R) = V$.

Proof. Let $\alpha_i: D^n \to M$, $i = 1,\dots,r$ be smooth embeddings of D^n into M such that $M = \bigcup_{i=1}^{r} \operatorname{int}(\alpha(D^n))$, where $\operatorname{int}(\alpha_i(D^n))$ is the interior of $\alpha_i(D^n)$, viewed as a subspace of M. Let $\psi_i: \varphi_i^*\xi \approx D^n \times R^q$ be a smooth trivialization of $\varphi_i^*\xi$. Define a linear map $\Psi : \mathcal{M}(\xi) \to \bigoplus_{i=1}^{r} \mathcal{M}(D^n,R^q)$ by $\Psi(s) = (s_1,\dots,s_r)$, where $s_i(x) = \psi_i(s(\varphi_i(x)))$. We may apply Lemma 1.19 to conclude that Ψ is an isomorphism of $\mathcal{M}(\xi)$ with $V = \{s_1,\dots,s_r) \in \bigoplus_{i=1}^{r} \mathcal{M}(D^n,R^q) \mid \psi_i^{-1}s_i\varphi_i^{-1} = \psi_j^{-1}s_j\varphi_j^{-1}\}$. But recall that $\mathcal{M}(D^n,R^q) = \bigoplus_{i=1}^{q} \mathcal{M}(D^n,R)$. Thus $\mathcal{M}(\xi)$ is completely determined by $\mathcal{M}(D^n,R)$.

Conversely, if V is a $\mathcal{C}$-space of functions on D^n satisfying (1) - (3) in the statement of the theorem, then it is obvious that we may use the isomorphism we derived in the first part of this proof to define an n-dimensional $\mathcal{C}$ section functor which extends V. □

Let $\mathcal{M}$ be an n-dimensional $\mathcal{C}$ section functor, and ξ a vector bundle over a compact n-dimensional manifold M. As a consequence of the above theorem, we can always find a commutative diagram of the following sort, with the left horizontal maps closed linear embeddings:

$$\begin{array}{ccccc}
\mathcal{M}(\xi) & \longrightarrow & \bigoplus_{i=1}^{r} \mathcal{M}(D^n,R^q) & = & \bigoplus_{i=1}^{qr} \mathcal{M}(D^n,R) \\
\downarrow & & \downarrow & & \downarrow \\
C^0(\xi) & \longrightarrow & \bigoplus_{i=1} C^0(D^n,R^q) & = & \bigoplus_{i=1}^{qr} C^0(D^n,R)
\end{array}$$

Thus certain properties will follow for the map $\mathcal{M}(\xi) \to C^0(\xi)$ if they are true for the map $\mathcal{M}(D^n,R) \to C^0(D^n,R)$. We will illustrate this point with Banach section functors for examples.

1.21 Definition. An n-dimensional Banach section functor is an n-dimensional $\mathcal{C}$ section functor to the CG category of Banachable spaces.

Let $\mathcal{M}$ be an n-dimensional Banach section functor. Returning to the above discussion, we see that, if the map $\mathcal{M}(D^n,R) \to C^0(D^n,R)$ is completely continuous, so is the map $\mathcal{M}(\xi) \to C^0(\xi)$. Thus it is possible to refer to $\mathcal{M}$ as a "completely continuous" Banach section functor.

Suppose $\mathcal{M}(D^n,R)$ is a reflexive space. Then $\mathcal{M}(\xi)$ is isomorphic to a closed subspace of the reflexive space $\bigoplus_{i=1}^{qr} \mathcal{M}(D^n,R)$, and so is also reflexive. If this is the case, we refer to $\mathcal{M}$ as a reflexive Banach section functor.

Incidentally, let $\mathcal{M}$ be a Banach section functor, and assume there exists a compact manifold M and vector bundle ξ over M such that the map $\mathcal{M}(\xi) \to C^0(\xi)$ is completely continuous. Choose an embedding of the disc $\varphi : D^n \to M$, and a trivialization of $\varphi^*\xi$. Then we have the following commutative diagram, with the left horizontal maps surjections:

$$\begin{array}{ccccccc}
\mathcal{M}(\xi) & \longrightarrow & \mathcal{M}(\xi|_{D^n}) & \approx & \mathcal{M}(D^n \times R^q) & = & \bigoplus_{i=1}^{q} \mathcal{M}(D^n,R) \\
\downarrow & & \downarrow & & \downarrow & & \downarrow \\
C^0(\xi) & \longrightarrow & C^0(\xi|_{D^n}) & \approx & C^0(D^n \times R^q) & = & \bigoplus_{i=1}^{q} C^0(D^n,R)
\end{array}$$

Thus the map $\mathcal{M}(\xi|_{D^n}) \to C^0(\xi|_{D^n})$ is completely continuous, which implies that the map $\mathcal{M}(D^n,R) \to C^0(D^n,R)$ is completely continuous. Similarly, $\mathcal{M}$ is reflexive $\iff$ there exists a compact manifold M and a vector bundle ξ over M such that $\mathcal{M}(\xi)$ is reflexive.

Let us consider for a moment a Banach section functor $\mathcal{M}$ which is both reflexive and completely continuous. Let ξ be a vector bundle over a compact n-dimensional manifold M, and choose a norm for $\mathcal{M}(\xi)$. Then the image of the closed unit ball of $\mathcal{M}(\xi)$, under the inclusion $\mathcal{M}(\xi) \to C^0(\xi)$, is a compact subset of $C^0(\xi)$.

More generally, let $f : V_1 \to V_2$ be a completely continuous linear map of the reflexive Banach space V_1 into the Banach space

V_2. Let $B(1)$ be the closed unit ball of V_1. Then $f(B(1))$ is a compact subset of V_2.

To see this, consider the commutative diagram,

$$\begin{array}{ccc} f : V_1 & \longrightarrow & V_2 \\ \downarrow & & \downarrow \\ f : V_1^w & \longrightarrow & V_2^w \end{array}$$

where V_i^w is V_i with the weak topology.

The unit ball $B(1)$ of V_1 is compact in V_1^w, so its image in V_2^w is compact. But $f(B(1))$ has compact closure in V_2, so the topology which $f(B(1))$ inherits as a subspace of V_2 is the same as its topology as a subspace of V_2^w. Thus $f(B(1))$ is compact in V_2. In addition, if f is injective (as it is for the map $\mathcal{M}(\xi) \to C^0(\xi)$), then f is a homeomorphism between $B(1)$ with the weak topology, and $f(B(1))$ with the V_2-topology.

This property that the unit ball of $\mathcal{M}(\xi)$ is compact in $C^0(\xi)$ is much more important than the reflexivity of $\mathcal{M}$. Thus it is natural to include in our discussion Banach section functors which have this property, but which are not reflexive.

Let $\mathcal{M}$ be an n-dimensional Banach section functor which is completely continuous but not reflexive. Then for each object ξ of FVB(n), there <u>always</u> exist norms on $\mathcal{M}(\xi)$ such that the unit balls in these norms are not compact in $C^0(\xi)$. So the natural question to ask is: does there exist <u>any</u> norm on $\mathcal{M}(\xi)$ for which the unit ball is compact in $C^0(\xi)$?

Assume there exists a norm on $\mathcal{M}(D^n, R)$ for which the image of the unit ball is compact. Then, by an argument identical to that used in the discussion of completely continuous functors above, it is easy to see that there exists a norm with the desired property on $\mathcal{M}(\xi)$ for each object ξ of FVB(n).

Conversely, if there exists an object ξ of FVB(n) such that $\mathcal{M}(\xi)$ has a compact norm, then there exists a compact norm on $\mathcal{M}(D^n, R)$.

1.22 Definition. An n-dimensional Banach section functor $\mathcal{M}$ is compact if there exists a norm on $\mathcal{M}(D^n, R)$ for which the unit ball of $\mathcal{M}(D^n, R)$ is compact in $C^0(D^n, R)$.

The following are the standard examples of n-dimensional Banach section functors which appear in global analysis:

1.23 Examples. (1) L^p_k, $p > 1$, $k > n/p$, is completely continuous and reflexive, and hence compact.

(2) $C^{k+\alpha}$, k a non-negative integer, $0 < \alpha < 1$, is compact, but not reflexive.

(3) C^{k-}, k a positive integer (the functions which are continuously differentiable $(k-1)$ times and whose $(k-1)$-st derivatives are all uniformly Lipschitz), is compact but not reflexive.

(4) C^k, k a positive integer, is completely continuous but not compact.

(5) L^1_1 is a one-dimensional Banach section functor, but is neither completely continuous nor reflexive.

(6) L^1_k, $k > n$, is an n-dimensional Banach section functor which is completely continuous but not compact.

(7) C^0 is a Banach section functor, but obviously has none of the properties we have been discussing.

It is possible to say something about section functors which are completely continuous and not compact, but the compact case is much easier to handle, and is the more important of the two cases. A discussion of completely continuous section functors which are not compact will be left to a future paper.

Let us consider the section functor C^0 for a moment. Let M be a compact manifold, and ξ a vector bundle over M. Since $C^0(\xi)$ is a Banachable space, we have a well-defined concept of bounded subsets of $C^0(\xi)$. Now let η be another vector bundle over M, and let $f \in \mathrm{Map}(\xi, \eta)$. Then $C^0(f)$ maps bounded subsets of $C^0(\xi)$ to bounded subsets of $C^0(\eta)$.

The surprising thing is that for all the examples of Banach section functors listed above, $\mathcal{M}(f)$ also sends bounded subsets of $\mathcal{M}(\xi)$ to bounded subsets of $\mathcal{M}(\eta)$. In fact, I do not know of any Banach section functor which does not have this property. As with the properties of complete continuity, compactness and reflexivity, for an n-dimensional Banach section functor to map bounded sets to bounded sets, it is necessary and sufficient that $\mathcal{M}(f) : \mathcal{M}(D^n, R^q) \to \mathcal{M}(D^n, R)$ send bounded sets to bounded sets for each $q \in N$, and $f \in \mathrm{Map}(D^n \times R^q, D^n \times R)$.

1.24 Definition. A Banach section functor $\mathcal{M}$ on FVB(M) is said to preserve boundedness if, for each pair ξ, η of vector bundles on M, and $f \in \mathrm{Map}(\xi, \eta)$, $\mathcal{M}(f) : \mathcal{M}(\xi) \to \mathcal{M}(\eta)$ sends bounded sets to bounded sets.

The fact that all the standard Banach section functors preserve boundedness was first noticed by R. Palais. K. Uhlenbeck developed and exploited this idea to establish existence of solutions to a large class of problems in the calculus of variations. We will also develop this concept, though in a somewhat different fashion from Mrs. Uhlenbeck's treatment.

For the remainder of this section, all Banach section functors are assumed to be compact. Throughout the rest of this paper, they are assumed to preserve boundedness.

It will often be useful in proofs to choose a norm for the Banachable space $\mathcal{M}(\xi)$. We will always assume that we have chosen a norm such that the unit ball of $\mathcal{M}(\xi)$ is compact in $C^0(\xi)$.

Let $\mathcal{M}$ be an n-dimensional Banach section functor (compact, and which preserves boundedness). It is possible to derive a section functor $\omega_{\mathcal{M}}$ (we will ignore the subscript and simply write ω) from $\mathcal{M}$ as follows: For each object ξ of FVB(n), let χ be the set of bounded subsets of $\mathcal{M}(\xi)$. Note that χ is partially ordered by inclusion (for $A, B \in \chi$, $A \leq B \iff A \subset B$).

For each $A \in \chi$, let $\widetilde{A}$ be a topological space with the same underlying set as A, but let the topology on $\widetilde{A}$ be the topology which A inherits as a subspace of $C^{o}(\xi)$, i.e. $\widetilde{A}$ has the same underlying set as A, but has the topology of C^{o} convergence instead of the topology of $\mathcal{M}$ convergence.

Define $\omega(\xi) = \varinjlim_{A \in \chi} \widetilde{A}$. Note that $\omega(\xi)$ has the same underlying set as $\mathcal{M}(\xi)$, and the topology on $\omega(\xi)$ is the strongest such that the inclusion $\widetilde{A} \to \omega(\xi)$ is continuous for all $A \in \chi$.

Observe that, since the map $A \to C^{o}(\xi)$ is continuous for all $A \in \chi$, the map $\omega(\xi) \to C^{o}(\xi)$ is continuous. Since we have a commutative diagram

$$\begin{array}{ccc} \widetilde{A} & \longrightarrow & \omega(\xi) \\ \cap & \swarrow & \\ C^{o}(\xi) & & \end{array},$$

the topology which $\widetilde{A}$ inherits as a subspace of $\omega(\xi)$ is the same as the topology which it inherits from $C^{o}(\xi)$.

1.25 Theorem. ω is a section functor on $FVB(n)$.

Proof. The underlying set of $\omega(\xi)$ is the same as the underlying set of $\mathcal{M}(\xi)$, and so consists of continuous sections of ξ. We observed above that the map $\omega(\xi) \to C^{o}(\xi)$ is continuous. Thus the only thing we must verify is that, for M a compact manifold, ξ and η vector bundles over M, and $f \in \mathrm{Map}(\xi,\eta)$, $\omega(f) : \omega(\xi) \to \omega(\eta)$ is continuous.

Since $\omega(\xi) = \varinjlim_{A \in \chi} \widetilde{A}$, it suffices to verify that $\omega(f) : \widetilde{A} \to \omega(\eta)$ is continuous for each $A \in \chi$. But A bounded in $\mathcal{M}(\xi) \Longrightarrow \mathcal{M}(f)(A)$ is bounded in $\mathcal{M}(\eta)$. Thus $\omega(f)(\widetilde{A})$ has the same topology as a subspace of $\omega(\eta)$ that it inherits as a subspace of $C^{o}(\eta)$, and so we have the following commutative diagram:

$$\begin{array}{ccc} \widetilde{A} & \subset & C^{o}(\xi) \\ \Big\downarrow \omega(f) & & \Big\downarrow C^{o}(f) \\ \omega(f)(\widetilde{A}) & \subset & C^{o}(\eta) \end{array}$$

Since $C^{o}(f)$ is continuous, $\omega(f) : \widetilde{A} \to \omega(f)(\widetilde{A}) \subset \omega(\eta)$ is continuous.

1.26 Corollary. ω extends to a unique section functor on FB(n).

Proof. Immediate from Theorem 1.9.

1.27 Corollary. Let E be a fiber bundle over a compact n-dimensional manifold M. The underlying set of $\omega(E)$ is the same as the underlying set of $\mathcal{M}(E)$.

Proof. Let Q be the set of open vector subbundles of E. Then $\omega(E) = \bigcup_{\xi \in Q} \omega(\xi)$, and $\mathcal{M}(E) = \bigcup_{\xi \in Q} \mathcal{M}(\xi)$. □

1.28 Corollary. For each open vector subbundle $\xi \subset E$, the induced inclusion $\omega(\xi) \to \omega(E)$ is a continuous embedding onto an open subset of $\omega(E)$.

Proof. This is Theorem 1.9 again. □

1.29 Corollary. The map $\mathcal{M}(E) \to \omega(E)$ is continuous, i.e. there exists a natural transformation $\mathcal{M} \to \omega$.

Proof. Let Q be the set of open vector subbundles of E. $\mathcal{M}(E) = \bigcup_{\xi \in Q} \mathcal{M}(\xi)$, and each $\mathcal{M}(\xi)$ is open in $\mathcal{M}(E)$. But the map $\mathcal{M}(\xi) \to \omega(\xi) \subset \omega(E)$ is continuous. □

1.30 Lemma. Let M be a compact manifold, ξ a vector bundle over M, and K a subset of $\mathcal{M}(\xi)$. Then K is bounded in $\mathcal{M}(\xi)$ <===> the closure of K in $\omega(\xi)$ is compact.

Proof. Assume K is bounded in $\mathcal{M}(\xi)$. Then there exists n such that $K \subset B(n)$, where B(n) is the closed ball of radius n in $\mathcal{M}(\xi)$. But $\mathcal{M}$ is compact, so B(n) is compact in $\omega(\xi)$, and hence also closed in $\omega(\xi)$. Thus the closure of K in $\omega(\xi)$ is a subset of B(n), and hence is compact.

Conversely, suppose K is not bounded in $\mathcal{M}(\xi)$. Choose inductively a sequence of elements s_n of K such that $\|s_n\| > n$, $s_n \notin \{s_1,\dots,s_{n-1}\}$. Let $B = \{s_n\}_{n\in N}$, and let C be any subset of B. Recalling the notation used in the definition of ω, observe that $C \cap \widetilde{A}$ is a finite set for each $\widetilde{A}$ such that $A \in \chi$ (χ = the set of bounded subsets of $\mathcal{M}(\xi)$). Thus C is a closed subset of $\widetilde{A}$ for every $A \in \chi$. Since $\omega(\xi) = \varinjlim_{\widetilde{A} \in \chi} \widetilde{A}$, C is closed in $\omega(\xi)$. Since C is closed in $\omega(\xi)$ for any subset C of B, B is a closed discrete subset of $\omega(\xi)$, and hence also of the closure of K in $\omega(\xi)$. But if the closure of K in $\omega(\xi)$ were compact, it could not have any infinite closed discrete subspaces. Thus K does not have compact closure in $\omega(\xi)$. □

1.31 Theorem. Let E be a fiber bundle over M, $i : \mathcal{M}(E) \to \omega(E)$, and $K \subset \mathcal{M}(E)$. Then $i(K)$ has compact closure in $\omega(E)$ <===> there exists a finite number of open vector subbundles $\xi_1, \ldots, \xi_r$ of E, and subsets $K_1, \ldots, K_r$ of $\mathcal{M}(E)$, such that K_j is a bounded subset of $\mathcal{M}(\xi_j)$ and $K = \bigcup_{j=1}^{r} K_j$.

Proof. Assume $K = \bigcup_{j=1}^{r} K_j$, K_j bounded in $\mathcal{M}(\xi_j)$. By the above lemma, the closure of K_j in $\omega(\xi_j)$ is compact, which implies that the closure of K_j in $\omega(\xi_j)$ coincides with $\overline{i(K_j)}$ (the closure of $i(K_j)$ in $\omega(E)$). Thus $\overline{i(K)} = \bigcup_{j=1}^{r} \overline{i(K_j)}$ is compact in $\omega(E)$.

To prove the implication in the other direction, assume $\overline{i(K)}$ is compact. Given $s \in \overline{i(K)}$, there exists an open vector subbundle ξ_s with $s \in \mathcal{M}(\xi_s)$. Choose an open vector subbundle η_s of E with $s \in \mathcal{M}(\eta_s)$ and $\overline{\eta}_s \subset \xi_s$. Then $\overline{C^0(\eta_s)} \subset C^0(\xi_s)$ (the closure is taken in $C^0(E)$), so $\overline{\omega(\eta_s)} \subset \omega(\xi_s)$. Note that the collection $\{\omega(\eta_s)\}_{s \in \overline{i(K)}}$ forms an open cover of $\overline{i(K)}$, so there exists a finite subcover $\omega(\eta_1), \ldots, \omega(\eta_r)$. Define $H_j \subset \mathcal{M}(E)$ by $i(H_j) = \overline{i(K)} \cap \overline{\omega(\eta_j)}$. Then $i(H_j)$ is a compact subset of $\omega(E)$ for each j. Since $i(H_j) \subset \omega(\xi_j)$, H_j is bounded in $\mathcal{M}(\xi_j)$ by Lemma 1.30. Finally, define $K_j = H_j \cap K$, $1 \le j \le r$. □

The observation that it is possible to give an intrinsic characterization of the notion of boundedness in manifolds of maps is due to K. Uhlenbeck in [45]. She defined the following notion of intrinsically bounded set:

1.32 Definition. $K \subseteq \mathcal{M}(E)$ is intrinsically bounded if:

(1) K is relatively compact in $C^0(E)$.

(2) For each subset $\tilde{K} \subset K$ and open vector subbundle ξ of E with $\tilde{K} \subseteq \mathcal{M}(\xi)$, $\tilde{K}$ is bounded in $\mathcal{M}(\xi)$ <==> $\tilde{K}$ is bounded in $C^0(\xi)$.

Her main theorem on boundedness states that K is intrinsically bounded <===> there exists a finite number of open vector subbundles $\xi_1,\ldots,\xi_r$ of E, and bounded subsets K_j of $\mathcal{M}(\xi_j)$, $1 \le j \le r$, such that $K = \bigcup_{j=1}^{r} K_j$. She then restricted her attention to the Sobolev manifolds $L_k^p(E)$, and showed the existence of a naturally-arising class of metrics on these manifolds for which the metrically bounded sets coincide with the intrinsically bounded sets. Her ideas have served to motivate an abstract treatment of metrics which will come later.

1.33 Remark. Assume that $\mathcal{M}$ is a reflexive section functor, and let ξ be a vector bundle over M. Recall that we showed, in the discussion which followed Definition 1.21, that the topology on $\tilde{A}$ for each $A \in \chi$ is the usual weak topology. Thus the topology on $\omega(\xi)$ is the topology generated by the weak topology on $\mathcal{M}(\xi)$-bounded sets. This topology is familiar to functional analysts under the name bounded weak topology. The bounded weak topology is locally convex, so that for any fiber bundle E, $\omega(E)$ is a topological manifold modeled on a locally convex space. The bounded-weak convergent sequences of a reflexive space are the same as the weakly convergent sequences, so that for any fiber bundle E, the concept of weakly convergent sequence in $\mathcal{M}(\xi)$ becomes a globally defined notion. The unit ball of $\mathcal{M}(\xi)$ with the weak topology is metrizable (it can be embedded in $C^0(\xi)$), so $\mathcal{M}(\xi)$ is separable. Since each bounded subset of $\mathcal{M}(\xi)$ with the weak topology is metrizable, the topology on each of these sets is sequentially generated. Since $\omega(\xi) = \varinjlim_{A \in \chi} \tilde{A}$, the topology on $\omega(\xi)$ is sequentially generated.

Since the $\{\omega(\xi),\ \xi$ is an open vector subbundle of $E\}$ forms an open cover of $\omega(E)$, the topology on $\omega(E)$ is sequentially generated, i.e. the topology on $\omega(E)$ is the topology generated by the weakly convergent sequences of $\mathcal{M}(E)$.

K. Uhlenbeck and J. Dowling independently proved, using analytical techniques, that for the n-dimensional Sobolev functor L^p_k, $k > n/p$, $L^p_k(f)$ preserves weakly convergent sequences for each morphism f of FVB(n). This fact was related to me without proof, and the material in this and the next chapter was motivated by an attempt to understand their result.

Since the bounded weak topology on the space $\mathcal{M}(\xi)$ is preserved by morphisms induced by fiber bundle mappings, it is natural to ask whether the weak topology is preserved. The answer is no, in general.

1.34 Example. Let M be a compact manifold, $\mathcal{M}$ a LTS section functor on FVB(M). Then the morphisms $\mathcal{M}(f)$, $f \in \mathrm{Map}(\xi,\eta)$ do not in general preserve the weak topology.

Proof. We may assume that M has a smooth measure. Consider the map $\varphi : M \times \mathbb{R} \to M \times \mathbb{R}$, $(x,y) \to (x,y^2)$. For each $s \in \mathcal{M}(M,\mathbb{R})$, $\mathcal{M}(\varphi)(s) = s^2$. Consider the continuous functional ℓ on $C^0(M,\mathbb{R})$ given by integration, i.e. $\ell(s) = \int_M s(x)dx$ for all $s \in C^0(M,\mathbb{R})$. Since the inclusion $\mathcal{M}(M,\mathbb{R}) \to C^0(M,\mathbb{R})$ is continuous, we may regard ℓ as a continuous linear functional on $\mathcal{M}(M,\mathbb{R})$. Since ℓ is continuous and linear, it is weakly continuous. Suppose for a moment that $\mathcal{M}(\varphi)$ were weakly continuous. Then $g = \ell \circ \mathcal{M}(\varphi)$ would be weakly continuous. Thus $g^{-1}((-1,1))$ would be weakly open in $\mathcal{M}(M,\mathbb{R})$. By definition of the weak topology, there would exist a finite number of linear functionals $f_1,\dots f_r$ such that $\bigcap_{i=1}^{r} f_i^{-1}(-1,1) \subset g^{-1}(-1,1)$. Since $C^\infty(M,\mathbb{R}) \subset \mathcal{M}(M,\mathbb{R})$, we know that $\mathcal{M}(M,\mathbb{R})$ is infinite dimensional. Thus $V = \bigcap_{i=1}^{r} f_i^{-1}(0)$ is a non-zero subspace of $\mathcal{M}(M,\mathbb{R})$. Choose a non-zero element $s \in V$. Since $s \in V$, $rs \in V$ for all $r \in \mathbb{R}$, $\Rightarrow |\ell\circ\mathcal{M}(\varphi)(rs)| \le 1$ for all $r \in \mathbb{R}$. But $\ell\circ\mathcal{M}(\varphi)(rs) = \int_M r^2(s(x))^2dx = r^2\int_M (s(x))^2dx$. Since s is not the zero function, $\int_M (s(x))^2 dx = a > 0$. But this implies that $\ell \circ \mathcal{M}(\varphi)(rs) > 1$ for all $r > 1/\sqrt{a}$. Contradiction. □

2. THE BOUNDED WEAK* TOPOLOGY

In the second half of Chapter 1 we discussed Banach section functors which preserve boundedness, and showed how to construct a new section functor ω from the compact Banach section functor $\mathfrak{M}$. If $\mathfrak{M}$ is reflexive, then $\omega(\xi)$ (ξ a vector bundle) is simply the space $\mathfrak{M}(\xi)$ with the bounded weak topology. For $\mathfrak{M}$ not reflexive, a reasonable conjecture would be that $\mathfrak{M}(\xi)$ is still the dual of some Banach space V, and that $\omega(\xi)$ is $\mathfrak{M}(\xi)$ with the bounded weak* topology. This conjecture, and much more, can be established with no increase in difficulty within the framework of abstract functional analysis.

<u>2.1 Definition.</u> Let V be a Hausdorff linear topological space, K a balanced compact subset of V. K is semi-convex if there exists $r > 0$ such that $x + y \in r \cdot K$ for all $x, y \in K$. The convexity number α_K of $K = \inf\{r \in R^+ | x + y \in rK$ for all $x, y \in K\}$.

<u>2.2 Remark</u>. Letting $y = x$, we see that $\alpha_K \geq 2$. It is easy to see that $\alpha_K = 2 \iff K$ is convex, but we do not need this result and so will not bother with the proof.

Assume that V is a linear topological space (all spaces in this paper are assumed to be Hausdorff), and K a semi-convex subset of V. We constrct a new topological space from V and K as follows:

<u>2.3 Definition</u>. $X = \varinjlim_{x \in R^+} rK = \varinjlim_{n \in N} nK$. We denote rK by K(r) for reasons which will become clear in Theorem 2.16. Thus $X = \varinjlim_{n \in N} K(n)$.

<u>2.4 Remark.</u> (1) X is a vector space.

(2) X is Hausdorff.

(3) X is compactly generated.

(4) The topology each $K(n)$ inherits as a subspace of X coincides with the topology which $K(n)$ inherits as a subspace of V.

Our first goal is to prove that X is a linear topological space. We shall need the following lemma, first proved by J. Milnor:

<u>2.5 Lemma</u>. Assume Y and Z are Hausdorff, $Y = \varinjlim_{n \in N} Y(n)$, $Z = \varinjlim_{n \in N} Z(n)$, $Y(n)$ and $Z(n)$ compact for all $n \in N$. Then $Y \times Z = \varinjlim_{n \in N} (Y(n) \times Z(n))$.

<u>Proof</u>. We have a continuous inclusion $Y(n) \times Z(n) \to Y \times Z$ for each $n \in N$, $\Longrightarrow$ the map $\varinjlim_{n \in N} (Y(n) \times Z(n)) \to Y \times Z$ is continuous. To complete the proof, it suffices to show that, for each open subset U of $\varinjlim_{n \in N} (Y(n) \times Z(n))$ and $(y,z) \in U$, there exists an open subset V of Y and an open subset W of Z such that $(y,z) \in V \times W \subset U$. To simplify notation, let $U(n) = U \cap (Y(n) \times Z(n))$, so that $U(n)$ is open in $Y(n) \times Z(n)$. We construct V and W as follows: choose N_o such that $(y,z) \in Y(N_o) \times Z(N_o)$. Choose $V(N_o)$ open in $Y(N_o)$ and $W(N_o)$ open in $Z(N_o)$ such that $(y,z) \in \overline{V(N_o)} \times \overline{W(N_o)} \subset U(N_o)$. Now let $n > N_o$, and assume that we have found, for each m such that $N_o < m < n$, open subsets $V(m)$ of $Y(m)$ and $W(m)$ of $Z(m)$ such that $\overline{V(m-1)} \times \overline{W(m-1)} \subset V(m) \times W(m) \subset \overline{V(m)} \times \overline{W(m)} \subset U(m)$. Since $\overline{V(n-1)} \subset Y(n-1)$, $\overline{V(n-1)}$ is compact. Similarly $\overline{W(n-1)}$ is compact. Since $\overline{V(n-1)} \times \overline{W(n-1)} \subset U(n)$ with $U(n)$ open in $Y(n) \times Z(n)$, there exists $V(n)$ open in $Y(n)$ and $W(n)$ open in $Z(n)$ such that $\overline{V(n-1)} \times \overline{W(n-1)} \subset V(n) \times W(n) \in \overline{V(n)} \times \overline{W(n)} \subset U(n)$. Thus by induction, we may find sequences $V(n)$ and $W(n)$ for all $n > N_o$ with the desired properties.

Define $V = \bigcup_{n \geq N_o} V(n)$, $W = \bigcup_{n \geq N_o} W(n)$.

Note that $V \cap Y(m) = \bigcup_{n \geq \max\{N_o, m\}} V(n) \cap Y(m)$. But $V(n) \cap Y(m)$ is an open subset of $Y(m)$ if $n \geq m$, $\Longrightarrow$ $V \cap Y(m)$ is open in $Y(m)$. Since this is true for all m, V is an open subset of Y. Similarly, W is an open subset of Z. And obviously $(y,z) \in V \times W \subset U$. □

2.6 Theorem. X is a linear topological space.

Proof. We must check that addition and scalar multiplication are continuous. By Lemma 2.5, $X \times X = \lim_{\substack{\rightarrow \\ n \in N}} (K(n) \times K(n))$.

Thus to check that addition is continuous, it suffices to check that add: $K(n) \times K(n) \to X$ is continuous. But $\text{add}(K(n) \times K(n)) \subset K(n\alpha_K)$. Since the sets $K(r)$, $r \in R^+$ have the same topology as subspaces of V that they have as subspaces of X, we have the following diagram:

$$\begin{array}{lccc} \text{add:} & K(n) \times K(n) & \longrightarrow & K(n\alpha_K) \subset X \\ & \cap & & \cap \\ \text{add:} & V \times V & \longrightarrow & V \end{array}$$

Thus the continuity of addition in X is a consequence of the continuity of addition in V.

The continuity of scalar multiplication follows similarly. □

If X is finite-dimensional, $\dim X = n$, then it is well known that X must be isomorphic to R^n. We shall now see that, if X is infinite-dimensional, X is not metrizable.

2.7 Proposition. Assume X is infinite-dimensional. Then X is not first countable.

Proof. Assume to the contrary that X is first countable. Then there exists a countable neighborhood basis $U_1, \ldots, U_n, \ldots$ at 0.

We may assume $U_{j+1} \subset U_j$. If any U_j were contained in some $K(n)$, then X would be locally compact, and hence finite-dimensional. Thus we may inductively choose a sequence s_n in X such that:

(1) $s_n \in U_n$

(2) $s_n \notin K_{n-1}$

Let $B = \{s_n\}_{n\in N}$. Since $B \cap K(n)$ is a finite set of points for each n, it follows by a standard argument that B is a closed discrete subset of X. But $s_n \in U_j$ for all $n \geq j$. Since $\{U_j\}_{j\in N}$ is a neighborhood basis of 0, $\lim_{n\to\infty} s_n = 0$. Thus we have obtained a contradiction, so there cannot exist a countable neighborhood basis of 0. □

Note that $K(n)$ is bounded in X for each $n \in N$. This is immediate since $K(n)$ is compact. Our next result is the converse:

2.8 Lemma. Assume A is a bounded subset of X. Then there exists N_o such that $A \subset K(N_o)$, i.e. $\bar{A}$ is compact.

Proof. We will assume that $K(n) - A \neq \emptyset$ for all $n \in N$, and show this implies that A is not bounded in X.

To demonstrate that A is not bounded, it suffices to construct an open neighborhood W of 0 such that αW does not contain A for any $\alpha \in R$. To see this, choose a sequence $x_n \in A$ such that $x_n \notin K(n2^n)$. Define $U_o = \{0\}$, $U_1 = K(1)$. Note that U_1 is an open subset of $K(1)$ (trivially), and that $x_n \notin 2^n \cdot U_1$ for all $n \in N$. Now let $j > 1$, and assume we have defined $U_i \subset K(i)$ for $1 \leq i < j$ such that:

(1) U_i is an open subset of $K(i)$

(2) $U_{i-1} \subset U_i$

(3) $x_n \notin 2^n U_i$ for all $n \in N$

For each $i \leq j$, let $\alpha_i : X \longrightarrow X$, $x \longrightarrow 2^i x$. Define a subset Y_j of X by

$Y_j = \bigcap_{i=1}^{j} \alpha_i^{-1}(X-\{x_i\})$. Since $X-\{x_i\}$ is open in X for each $i \le j$, Y_j is open in X. Note also that $U_{j-1} \subset Y_j$. Now define $U_j = Y_j \cap K(j)$. Then U_j is open in $K(j)$, $U_{j-1} \subset U_j$, and $x_i \notin 2^i U_j$ for all $i \le j$. But if $i > j$, then $2^i U_j \subset 2^i K(j) = K(2^i \cdot j) \subset K(2^i \cdot i)$. Thus $x_i \notin 2^i U_j$ for all $i > j$, $\Rightarrow$ $x_n \notin 2^n \cdot U_j$ for all $n \in N$. By induction, there exists a sequence of sets $\{U_i | i \in N\}$ with the three desired properties. Define $U = \bigcup_{i \in N} U_i$. Then it is immediate that U is an open neighborhood of 0 in X, and $x_n \notin 2^n U$ for all $n \in N$.

To finish the proof, let W be a nonempty balanced open subset of U. Since $(-\alpha)W = \alpha W$, it suffices to show that $\alpha W - A \neq \emptyset$ for all $\alpha > 0$. So assume $\alpha > 0$, and choose n such that $2^n > \alpha$. Then $x_n \notin 2^n U$, so $x_n \notin 2^n W$. But since W is balanced, $\alpha W \subset 2^n W$, so $x_n \notin \alpha W$. □

<u>2.9 Remark</u>. X is quasi-complete.

<u>Proof</u>. Since closed and bounded subsets of X are compact, this is immediate. □

<u>2.10 Proposition</u>. Assume X is locally convex. Then there exists a subset $C \subset X$ such that C is compact and convex, and $X = \varinjlim_{n \in N} C(n)$.

<u>Proof</u>. Define C to be the closed convex hull of K. Since X is locally convex, the convex hull of any bounded subset of X is also bounded. Thus C is closed and bounded, and hence is compact by Lemma 2.8. But Lemma 2.8 also implies there exists N_o such that $C \subset K(N_o)$. Thus $K \subset C \subset K(N_o)$, which implies that $X = \varinjlim_{n \in N} C(n)$. □

Our next result is essentially the converse of Lemma 2.10:

2.11 Theorem. Assume that V is locally convex and K is convex. Then X is locally convex.

Proof. Assume W is an open neighborhood of 0 in X. We must construct a convex balanced open neighborhood U of 0 such that $U \subset W$. Define $U_o = \{0\}$, and let $j > 0$. Assume U_i has been defined for $0 < i < j$ such that:

(1) $U_{i-1} \subset U_i$

(2) U_i is open in $K(i)$

(3) $\overline{U}_i \subset W$

(4) U_i is balanced and convex

By definition of the topology on X, there exists an open subset W_j of V such that $W \cap K(j) = W_j \cap K(j)$. Choose Y_j open in V such that $\overline{U}_{j-1} \subset Y_j \subset \overline{Y}_j \subset W_j$. Addition is continuous in V, so $\mathrm{add}^{-1}(Y_j)$ is open in $V \times V$. Since $\{0\} \times \overline{U_{j-1}}$ is compact, and $\{0\} \times \overline{U_{j-1}} \subset \mathrm{add}^{-1}(Y_j)$, the local convexity of V implies there exists an open convex balanced subset G_j of V such that:

(1) $0 \in G_j$

(2) $G_j \times \overline{U_{j-1}} \subset \mathrm{add}^{-1}(Y_j)$

Define $Z_j = G_j + \overline{U_{j-1}}$. Then Z_j is convex and balanced, and is open in V. Note also that $Z_j \subset Y_j$.

Define $U_j = Z_j \cap K(j)$. Obviously U_j is open in $K(j)$, and U_j is convex and balanced since both Z_j and $K(j)$ are convex and balanced. Finally, $\overline{U}_j \subset \overline{Z}_j \subset \overline{Y}_j \subset W_j$.

By induction, there exists a sequence $\{U_i \mid i \in N\}$ of subsets of X with properties (1) - (4) above. Define $U = \bigcup_{i\in N} U_i$. Then U is convex, balanced and open, and $0 \in U \subset W$. □

From now on, we will assume that K is convex and V is locally convex. Thus X is locally convex, and it is possible to obtain a very explicit characterization of the topology on X in terms of certain Banach spaces which are naturally associated with X:

specifically, X' and X". Recall that the topology on X' is defined by the set of semi-norms $\{\lambda_M \mid M$ is a bounded subset of $X\}$, where $\lambda_M(f) = \sup\{|f(x)| \mid x \in M\}$ for all $f \in X'$. Since we have a distinguished subset K of X, it follows that we have a distinguished semi-norm λ_K on X'. Note that $\lambda_K(f) = 0 \Longleftrightarrow f = 0$. Thus λ_K is actually a norm on X'.

2.12 Lemma. The topology induced on X' by the set of semi-norms $\{\lambda_M \mid M$ is a bounded subset of $M\}$ coincides with the topology induced on X' by the norm λ_K. Furthermore, X' is complete, so that X' is a Banach space with distinguish norm λ_K.

Proof. To verify the first statement, it suffices to show that, for each bounded subset M of X, there exists $n_M \in N$ such that $\lambda_M(f) \leq n_M \lambda_K(f)$ for all $f \in X'$. Now, by Lemma 2.8, there exists $n_M \in N$ such that $M \subset n_M K$. Thus

$\lambda_n(f) = \sup\{|f(x)| \mid x \in M\} \leq \sup\{|f(x)| \mid x \in n_M K\}$

$= n_M\{|f(x)| \mid x \in K\} = n_M \lambda_K(f)$. Thus X' is a normed space.

To see that X' is complete, let $\{f_n\}_{n \in N}$ be a Cauchy sequence in X'. Then, for each $x \in X'$, $|f_n(x) - f_m(x)| \to 0$ as $n,m \to \infty$. Furthermore, $\sup\{|f_n(x) - f_m(x)| \mid x \in K\} \to 0$ as $n,m \to \infty$. Define a map $f : X \to R$ as follows: $f(x) = \lim_{n\to\infty} f_n(x)$. The map f is obviously linear; we must show that it is also continuous. But $\lim_{n\to\infty} \sup\{|f(x) - f_n(x)| \mid x \in K\} = 0$, so that $f|_K$ is the uniform limit of a sequence of continuous functions on K. Thus f is continuous on K. Since f is linear, f is continuous on nK for all $n \in N$. But $X = \lim\limits_{\substack{\to \\ n \in N}} nK$, so that f is continuous on X. Since $\lambda_K(f - f_n) = \sup\{|f(x) - f_n(x)| \mid x \in K\}$, $\lim_{n\to\infty} \lambda_K(f - f_n) = 0$, i.e. X' is complete. □

From now on, we will write $\|f\|$ for $\lambda_K(f)$.

Let V be a locally convex space. Recall that there exists a canonical injective linear map $J_V : V \to V''$, defined by

$(J_V(v))(f) = f(v)$ for all $f \in V'$, $v \in V$. This map in general is neither continuous nor onto.

2.13 Definition. A locally convex space V is semi-reflexive if the map J_V is onto.

2.14 Remark. Assume V is semi-reflexive, so that we have a canonical identification of V with V''. Then the weak topology on V coincides with the weak* topology on V''.

To make use of the concept of semi-reflexivity, we need the following standard result:

2.15 Lemma. A locally convex linear topological space V is semi-reflexive $\Longleftrightarrow$ every closed, convex, balanced and bounded subset of V is compact in the weak topology of V.

Proof. See [47], p. 140. □

2.16 Theorem. Let $E = X''$. Then:

(1) E is a Banach space with the same underlying vector space as X.

(2) The closed unit ball of E coincides as a set with K.

(3) A subset $A \subseteq E$ is bounded in E $\Longleftrightarrow$ it is bounded in X.

(4) The topology on X is the bounded weak* topology on E.

(5) The canonical linear map $E \to X$ is continuous.

Proof. Since bounded subsets of X have compact closure, (1) follows from Lemma 2.15, and the observation that the dual of X' is a Banach space. To prove (2), let $f \in X'$, and note that $\|f\| \le 1 \Longleftrightarrow |f(x)| \le 1$ for all $x \in K$. Thus the unit ball of X' consists of $\{f \in X' \mid |f(x)| \le 1 \text{ for all } x \in K\}$. Now, for each $x \in E$, $\|x\|_E \le 1 \Longleftrightarrow |f(x)| \le 1$ for all $f \in X'$ with $\|f\| \le 1$.

Suppose $x \in K$. Then $|f(x)| \le 1$ for all f in the unit ball of X', so $\|x\|_E \le 1$. Conversely, suppose $x \notin K$. Then by Mazur's Theorem (see [47], p. 108), there exists $f_o \in X'$ such that

$f_o(x) > 1$ and $|f_o(y)| \le 1$ for all $y \in K$. But since $|f_o(y)| \le 1$ for all $y \in K$, $\|f_o\| \le 1$. Thus $\|x\|_E \ge |f_o(x)| > 1$, so K coincides as a set with the unit ball of E.

To see (3), let $A \subset E$ and recall that, by definition, A is bounded in $E \iff A$ is absorbed by the unit ball of E. But Lemma 2.8 implies that A is bounded in $X \iff A$ is absorbed by K.

To prove (4), note that Remark 2.14 implies that the weak* topology on E coincides with the weak topology on X. Since the bounded subsets of E are just the bounded subsets of X, the bounded weak* topology on E is the bounded weak topology on X. Since each bounded subset of X is absorbed by K, the bounded weak topology on X is just $\varinjlim_{n\in N} K(n)$, where $K(n)$ is $K(n)$ with the weak topology. But $K(n)$ is a compact subset of X, so that the weak topology on $K(n)$ coincides with the original topology. Thus X with the bounded weak topology equals $\varinjlim_{n\in N} K(n)$, which is again X. Finally, (5) follows by simply noting that the bounded weak* topology on E is weaker than the norm topology. □

Recall the following definition:

2.17 Definition. A subset A of the locally convex space V is called a barrel if A is a convex, balanced, and absorbing closed subset of V.

Let A be a compact barrel in X. Then A is a barrel in E. But it is a standard result (since Banach spaces are of the second Baire category) that barrels in Banach spaces are neighborhoods of zero, so there exists $r > 0$ such that $K(r) \subset A$. Also, since A is compact, there exists $s > 0$ such that $A \subset K(s)$. This implies that $X = \varinjlim_{n\in N} A(n)$.

2.18 Corollary. There is a natural one-to-one correspondence between the following sets:

(1) Compact barrels in X.

(2) Norms on X'.

(3) Dual norms on E.

2.19 Definition. A locally convex space V is a bw* space if:

(1) There exists a compact barrel of V which absorbs all compact subsets of V.

(2) V is compactly generated.

2.20 Example. Let B be a Banach space, and let $E = B'$. Then E with the bounded weak* topology is a bw* space. In particular, by Theorem 2.16, X is a bw* space.

2.21 Lemma. Let V be a bw* space, K a compact barrel of V. Then $V = \varinjlim_{n \in N} K(n)$.

Proof. Let $\chi = \{C \subset V : C \text{ is compact}\}$, and let K_1 be a compact barrel in V which absorbs each $C \in \chi$. Since V is compactly generated, $V = \varinjlim_{C \in \chi} C$. But for each $C \in \chi$, there exists $n_C \in N$ such that $C \subset K_1(n_C)$. Thus $V = \varinjlim_{n \in N} K_1(n)$, and so V is an example of the direct limit spaces we have been considering in this chapter. Thus the remarks following 2.17 imply that $V = \varinjlim_{n \in N} K(n)$. □

Thus we conclude that the class of spaces included in our construction at the beginning of this chapter was exactly the class of bw* spaces, with one difference: the space X has a distinguished barrel, whereas a bw* space does not. Thus the dual of a bw* space V is a Banachable space, and the second dual of V, while still canonically identified with V, is also a Banachable space. Note that, by Corollary 2.18, the choice of a compact barrel in V is equivalent to the choice of a dual norm on V'. From now on we shall deal with bw* spaces, and shall use X to denote a bw* space. Whenever it is convenient to have a norm on $E = X''$, we shall choose a fixed compact barrel of X, and let E have the associated dual norm.

It will be useful to have an explicit set of semi-norms which define the topology on X:

2.22 Lemma. Let X be a bw* space, and let $M = \{A \mid A = \{f_i\}_{i\in N}$ is a sequence of elements of X' which converges to $0\}$. Then the semi-norms $\{\lambda_A \mid A \in M\}$ are continuous on X, and generate the bw* topology on X.

Proof. See [8], p. 427. □

The following result is taken from [37]:

2.23 Proposition. X is a complete linear topological space.

Proof. Let $\{x_\lambda \mid \lambda \in \Lambda\}$ be a Cauchy net in X. Then, for each $f \in X'$, $\{f(x_\lambda) \mid \lambda \in \Lambda\}$ is a Cauchy net of real numbers. Define a map $\alpha : X' \to R$ by $\alpha(f) = \lim\limits_{\substack{\to \\ \lambda\in\Lambda}} f(x_\lambda)$ for each $f \in X'$. The map α is obviously linear; we must now show it is continuous.

Since X' is a Banach space, α is continuous $\iff$ α is a bounded linear functional. Assume for a moment that α is not continuous. Then there exists a sequence $g_n \in X'$ such that $\|g_n\| = 1$, $\alpha(g_n) > n^2$. Let $f_n = g_n/n$. Then $\lim\limits_{n\to\infty} f_n = 0$, and $\alpha(f_n) \to \infty$. Put $S = \{f_n\}_{n\in N}$, and apply Lemma 2.22. Since $\{x_\lambda\}_{\lambda\in\Lambda}$ is a Cauchy net, there exists $\lambda_o \in \Lambda$ such that, for all $\beta, \gamma \geq \lambda_o$, $\lambda_S(x_\beta - x_\gamma) < 1 \Rightarrow |f_i(x_\beta - x_{\lambda_o})| < 1$ for all $\beta \geq \lambda_o$ and for all $i \in N$. But this implies $|\alpha(f_i) - f_i(x_{\lambda_o})| \leq 1$ for all $i \in N$, $\Rightarrow \{\alpha(f_i)\}_{i\in N}$ is a bounded sequence of numbers. Since this contradicts our assumption that α is not continuous, α must be continuous, hence $\alpha \in X$. And obviously $\lim\limits_{\substack{\to \\ \lambda\in\Lambda}} \lambda_A(\alpha - x_\lambda) \to 0$ for each sequence $A = \{g_i\}_{i \in N}$ in X' such that $\lim\limits_{i\to\infty} g_i = 0$. □

<u>2.24 Closed Graph Theorem.</u> Let X_1, X_2 be bw* spaces, $f : X_1 \to X_2$ a continuous linear bijection. Then f is an isomorphism.

<u>Proof</u>. Let K_1 be a barrel in X_1, and let $K_2 = f(K_1)$, so that K_2 is a compact convex balanced subset of X_2. To show that K_2 is a barrel in X_2, we must show that K_2 is absorbing. So consider $f : E_1 \to E_2$, where $E_1 = X_i''$. Since K_2 is compact, it is bounded in E_2, so that f is a bounded linear map of E_1 into E_2, hence is continuous. By the closed graph theorem for Banach spaces, $f(K_1)$ is a neighborhood of 0 in E_2, and hence is an absorbing subset of E_2. Since the underlying Linear spaces of E_2 and X_2 are the same, K_2 is also absorbing in X_2. Thus, by the remarks following 2.17, we see that $X_2 = \lim\limits_{\substack{\to \\ n\in N}} K_2(n)$.

Now, K_1 compact $\Longrightarrow$ $f : K_1 \to K_2$ is a homeomorphism, $\Longrightarrow f^{-1} : K_2 \to X_1$ is continuous, $\Longrightarrow f^{-1} : X_2 \to X_1$ is continuous. □

<u>2.25 Proposition</u>. Let X, X_1 be bw* spaces, and let Z be a closed linear subspace of X. Then $X \times X_1$, Z, and X/Z are bw* spaces.

<u>Proof</u>. Let $X = \lim\limits_{\substack{\to \\ n\in N}} K(n)$, $X_1 = \lim\limits_{\substack{\to \\ n\in N}} K_1(n)$. Then

$X \times X_1 = \lim\limits_{\substack{\to \\ n\in N}} (K\times K_1)(n)$ by Lemma 2.5, so that $X \times X_1$ is a bw* space.

Since Z is a closed subspace of X, it is compactly generated. Let $C = K \cap Z$, so that C is a compact, convex, balanced subset of Z. Let A be any compact subset of Z. Then there exists $n \in N$ such that $A \subset K(n)$, $\Longrightarrow A \subset K(n) \cap Z = C(n)$, $\Longrightarrow Z = \lim\limits_{\substack{\to \\ n\in N}} C(n)$,

$\Longrightarrow Z$ is a bw* space.

To see that X/Z is a bw* space, let $p: X \to X/Z$, and let

$C = p(K)$. Note that C is a compact, convex and balanced subset of X/Z, and that $X/Z = \bigcup_{n\in N} C(n)$. Define $X_2 = \varinjlim_{n\in N} C(n)$.

Then X_2 is a bw* space with the same underlying vector space as X/Z, and the map $X_2 \to X/Z$ is continuous. Since $p(K(n)) \subset C(n)$, and since $X = \varinjlim_{n\in N} K(n)$, the map $p: X \to X_2$ is also continuous.

But the topology on X/Z is the strongest topology such that the map p is continuous, $\Longrightarrow X_2 = X/Z$. □

Let us return for a moment to the bw* space X constructed in Definition 2.3. Assume that the locally convex space V used in the construction is a Banach space. Then K is a compact metric space.

2.26 Lemma. Let X be a bw* space, K a compact barrel in X. Then K is metrizable $\Longleftrightarrow X'$ is separable.

Proof. See [8], p. 426. □

2.27 Definition. A Bw* space X is a bw* space such that X' is separable.

2.28 Proposition. Let X, X_1 be Bw* spaces, Z a closed subspace of X. Then $X \times X_1$, Z, and X/Z are Bw* spaces.

Proof. Recalling Lemma 2.25, it is clear that we need only show that $(X\times X_1)'$, Z', and $(X/Z)'$ are separable. Since $(X\times X_1)' = X' \times X_1'$, it is separable. For the other two spaces, simply note that the exact sequence $0 \to Z \to X \to X/Z \to 0$ induces a dual exact sequence of Banach spaces $0 \to (X/Z)' \to X' \to Z' \to 0$. Since X' is separable, so are $(X/Z)'$ and Z'. □

2.29 Remark. The category of bw* spaces and the category of Bw* spaces are both examples of the CG categories of Definition 1.17.

2.30 Remark. (1) Let $n \in N$, and let η be an n-dimensional bw* section functor. Assume ξ is an object of $FVB(n)$, and let K

be a compact barrel of $\mathcal{N}(\xi)$. The natural injection $i : \mathcal{N}(\xi) \to C^0(\xi)$ induces a homeomorphism of K with $i(K)$. Thus K is metrizable, $\Longrightarrow$ $\mathcal{N}(\xi)$ is a Bw* space, $\Longrightarrow$ all bw* section functors are actually Bw* section functors.

(2) Let $\mathcal{M}$ be a compact n-dimensional Banach section functor. Then the section functor ω on $FVB(n)$ which we constructed from $\mathcal{M}$ in Chapter 1 is an n-dimensional Bw* section functor.

If X_1 and X_2 are bw* spaces, $f : X_1 \to X_2$ a continuous linear map, then f maps bounded sets to bounded sets, so that $f : E_1 \to E_2$ is continuous. Thus we may regard J in Definition 2.13 as a functor from the category of bw* spaces to the category of Banachable spaces, $J(X_i) = E_i$, $J(f) = f$.

Natural questions to ask are:

(1) Does the underlying linear space of every Banachable space E admit the structure of a bw* space X such that $X'' = E$?

(2) If E admits such a bw* structure, is it unique?

(3) If $f : X_1 \to X_2$ is a linear map such that f is a continuous map from E_1 to E_2, is f a continuous map from X_1 to X_2?

If we restrict our consideration to the category of reflexive Banachable spaces, then the answers to (1) - (3) are all yes. The general case is not quite so simple. I do not know the answer to (1), though I doubt it to be true. (2) is false, i.e. there exists a Banachable space E which admits two distinct Bw* topologies X_1 and X_2. Then, considering the canonical map $i : X_1 \to X_2$ which is the identity on the underlying vector spaces of X_1 and X_2, i is not continuous from X_1 to X_2 but $i : X_1'' \to X_2''$ is continuous, since $X_1'' = X_2'' = E$. The following is a well-known example of a Banach space which admits two distinct Bw* topologies:

<u>2.31 Example</u>. Let c_0 be the Banach space of sequences of real numbers $a = (a_1, a_2, \ldots)$ such that

$\lim_{n\to\infty} a_n = 0$, with the norm defined by $\|a\| = \sup\{|a_n| \mid n \in N\}$.

Similarly, let c_1 be the space of bounded sequences $b = (b_1, b_2, \ldots)$ such that $\lim_{n\to\infty} b_n$ exists, with $\|b\| = \sup\{|b_n| \mid n \in N\}$. Finally, let ℓ_1 be the space of absolutely summable sequences $e = (e_1, e_2, \ldots)$, with $\|e\| = \sum_{i=1}^{\infty} |e_i|$. It is easy to see that ℓ_1 has a natural identification with the dual of c_0 and also with the dual of c_1. In each case the induced dual norm on ℓ_1 is the original ℓ_1-norm, but the weak* topologies (and hence the bounded weak* topologies) are distinct.

Note that the closed unit ball of ℓ_1 is a compact barrel in both bw* topologies. Also, both c_0 and c_1 are separable, so that the two weak structures are Bw* topologies.

Convergent sequences are particularly interesting in bw* spaces. Let X be a bw* space, and let $s_n \to s$ be a bw*-convergent sequence. Since the bw* topology on X is stronger than the w* topology, the sequence s_n is w*-convergent. Conversely, let $s_n \to s$ be a a w*-convergent sequence in X. Regarding the elements of X as linear functionals on X', the uniform boundedness principle for Banach spaces implies that $\{s_n \mid n \in N\}$ is bounded in E, and hence in X. Thus the sequence s_n is bw*-convergent, and we conclude that a sequence s_n in X is w*-convergent $\Longleftrightarrow$ it is bw*-convergent.

<u>2.32 Lemma</u>. Let X be a Bw* space. Then the topology on X is sequentially generated.

<u>Proof</u>. Since X is a bw*-space, X is compactly generated. But X is a Bw* space, which implies that all compact subsets of X are metrizable, hence sequentially generated. The result is now immediate. □

Recall our original construction of a bw* space X from a locally convex space V and a compact, convex, balanced subset K of V. We have already remarked that, if V is a Banach space, X is a Bw* space. A converse of this fact is an obvious consequence of our next result:

2.33 Lemma. Let X be a Bw* space. Then there exists a continuous dense linear injection $i : X \to H$, where H is a separable Hilbert space.

Proof. Choose a sequence $\{f_n | n \in N\}$ of elements of unit norm in X' such that $\{f_n | n \in N\}$ is a dense subset of the unit sphere of X'. Define a pre-Hilbert structure on X by

$$\langle y,z \rangle_H = \Sigma_{n=1}^{\infty} \frac{1}{2^n} f_n(y)\, f_n(z) \quad \text{for all } y, z \in X.$$

Since $\{f_n | n \in N\}$ is dense in the unit sphere of X', $\langle x,x \rangle = 0 \iff x = 0$. Let H be the completion of X in this norm. Since X is sequentially generated, to check the continuity of the map $i : X \to H$, it suffices to check that $\|x_n\|_H \to 0$ for each sequence x_n in X with $\lim\limits_{n\to\infty} x_n = 0$. But this is obvious. To see that H is separable, let $X = \lim\limits_{\substack{\to \\ n\in N}} K(n)$, and note that $\bigcup\limits_{n\in N} i(K(n))$ is dense in H. Since each $i(K(n))$ is compact, this implies that H is separable. □

We already know, by definition, essentially, that bw* spaces are σ-compact. This will be an extremely important property when we discuss weak manifolds, and can be established for all open subsets of Bw* spaces. But first, we present a standard result about compactly generated spaces:

2.34 Lemma. Let Y be a compactly generated topological space, and assume in addition that Y is regular. Then every open subspace of Y is compactly generated.

Proof. Let U be an open subset of Y, and let A be a subset of U which has the property that $A \cap C$ is compact for each compact subset C in U. We will show that A is a closed subset of U by showing that $A \cup U'$ is a closed subset of Y.

To see that $A \cup U'$ is closed in Y, it suffices to show that $(A \cup U') \cap C$ is a closed subset of Y for each compact subset C of Y. So let C be a compact subset of Y: we will show that

$(A \cup U') \cap C$ is closed by showing that it contains all of its limit points. Let z be a limit point of $(A \cup U') \cap C$. Then z is a limit point of $A \cup U'$ and z is a limit point of C. Since C is compact, z must be contained in C. So it suffices to show that z is contained in $A \cup U'$.

Now, since z is a limit point of $A \cup U'$, z must be a limit point of either A or U'. If z is a limit point of U' then $z \in U'$, since U' is closed in Y. So we are reduced to considering the case where z is a limit point of A and $z \in U$. Since Y is regular, there exists an open subset W of Y such that $z \in W \subset \overline{W} \subset U$. Since z is a limit point of $(A \cup U') \cap C$, z is also a limit point of $(A \cup U') \cap C \cap \overline{W}$. But $(A \cup U') \cap C \cap \overline{W} = A \cap C \cap \overline{W}$. Now, $C \cap \overline{W}$ is a compact subset of U. Thus, by assumption, $A \cap C \cap \overline{W}$ is closed in U. Since $z \in U$ and z is a limit point of $A \cap C \cap \overline{W}$, it follows that $z \in A \cap C \cap \overline{W}$, and hence that $z \in A \cup U'$. □

<u>2.35 Remark.</u> Since all linear topological spaces are regular, the above lemma implies in particular that open subsets of bw* spaces are compactly generated.

<u>2.36 Notation.</u> Let X be a bw* space. The two topologies on X of interest to us are the bounded weak* topology and the associated Banach space topology. So let A be a subset of X. By the <u>weak topology on A</u> we shall mean the topology which A inherits as a subspace of X. By the <u>strong topology on A,</u> we mean the topology which A inherits as a subspace of E.

<u>2.37 Lemma.</u> Let V be a locally convex space, U an open convex subset of V, and C a compact subset of U. Then the closed convex hull of C is contained in U.

<u>Proof</u>. Straightforward. □

<u>2.38 Lemma</u>. Let X be a bw* space, λ a continuous semi-norm on X, and $U = \lambda^{-1}([0,1))$. Then there exists an increasing

sequence C_n of non-empty compact convex balanced subsets of $U \ni U = \varinjlim_{n\in N} C_n$.

Proof. Let K be a compact barrel in X, so that $X = \varinjlim_{n\in N} K(n)$. Define $C_n = K(n) \cap \lambda^{-1}([0,1-1/n])$. Then C_n is compact, convex and balanced, $C_n \subset C_{n+1}$ for each $n \in N$, and $\bigcup_{n\in N} C_n = U$. Recall that, by 2.34, U is compactly generated. Thus to show that $U = \varinjlim_{n\in N} C_n$, it suffices to show that, for each compact subset A in U, there exists $n_A \in N$ such that $A \subset C_{n_A}$. So let A be a compact subset of U. Then A is a compact subset of X, and hence is bounded in X. Thus there exists $n_o \in N$ such that $A \subset K(n_o)$. Now, for each $j \in N$, let $W_j = K(n_o) \cap \lambda^{-1}([0,1-1/j))$. Then $\{W_j\}_{j\in N}$ is an increasing sequence of open subsets of $K(n_o)$ and, since $\bigcup_{j\in N} W_j = K(n_o) \cap U$, $\{W_j\}_{j\in N}$ is an open cover of A. Since A is compact, there exists $j_o \in N$ such that $A \subset W_{j_o}$. So define $n_A = \max\{n_o, j_o\}$. Then

$$A \subset W_{j_o} = K(n_o) \cap \lambda^{-1}([0,1-1/j_o)) \subseteq K(n_A) \cap \lambda^{-1}([0,1-1/n_A]) = C_{n_A}. \quad \square$$

2.39 Theorem. Let X be a bw* space, U a non-empty open σ-compact subset of X. Then there exists an increasing sequence C_n of non-empty compact subsets of U such that $U = \varinjlim_{n\in N} C_n$. If U is convex, the C_n may be chosen to be convex.

Proof. For each $x \in U$, choose a continuous semi-norm λ_x on X such that $x+\lambda_x^{-1}([0,1)) \subset U$. Let $W_x = x+\lambda_x^{-1}([0,1))$. Then $\{W_x : x \in U\}$ is an open cover of U. Since U is σ-compact, this cover has a countable subcover $\{W_i\}_{i\in N}$. The previous lemma implies that, for each $i \in N$, there exists an increasing sequence $\{C_{ij}\}_{j\in N}$ of compact subsets of W_i such that $W_i = \varinjlim_{j\in N} C_{ij}$. For each $n \in N$, let $C_n = \bigcup_{\substack{i<n \\ j<n}} C_{ij}$. Then C_n is an increasing sequence of compact subsets of U, and $\bigcup_{n\in N} C_n = \bigcup_{i,j\in N} C_{ij} = U$. To see that $U = \varinjlim_{n\in N} C_n$, let A be a subset of U such that $A \cap C_n$ is an open subset of C_n for each $n \in N$. Then, for each pair $i,j \in N$, $A \cap C_{ij}$ is an open subset of C_{ij}. Since $W_i = \varinjlim_{j\in N} C_{ij}$, this implies that $A \cap W_i$ is open in W_i for each $i \in N$. But $U = \varinjlim_{i\in N} W_i$, and so we conclude that A is open in U. Thus $U = \varinjlim_{n\in N} C_n$.

If U is convex, let D_n be the closed convex hull of C_n for each $n \in N$. Then $C_n \subset D_n \subset U$ for each $n \in N$, and the D_n form an increasing sequence of sets. Thus $U = \varinjlim_{n\in N} D_n$. Since each C_n is bounded, each D_n is closed and bounded, and hence compact. And, of course, each D_n is also convex. □

2.40 Lemma. Lex X be a bw* space, U an open subspace of X which is σ-compact. Then U is paracompact.

Proof. It is standard point-set topology that σ-compact regular spaces are paracompact. □

2.41 Lemma. Lex X be a Bw* space, U an open subspace of X. Then U is σ-compact.

Proof. Let K be a compact barrel in X, so that $X = \varinjlim_{n\in N} K(n)$. For each $n \in N$, $U \cap K(n)$ is an open subspace of $K(n)$. But open subsets of compact metric spaces are σ-compact, so $U \cap K(n)$ is σ-compact for each $n \in N$. Thus $U = \bigcup_{n\in N} (U \cap K(n))$ is σ-compact. □

<u>2.42 Theorem</u>. Let X be a Bw* space, U a non-empty subspace of X. Then U is paracompact. Furthermore, there is an increasing sequence C_n of non-empty compact subsets of U such that $X = \varinjlim_{n \in N} C_n$. If U is convex, the C_n may be chosen to be convex.

<u>Proof</u>. Follows from 2.39, 2.40, and 2.41. □

Let X be a bw* space whose dual space X' is not separable, and let U be a neighborhood of 0. It is shown in [14] that there is an open neighborhood W of 0 such that $W \subseteq U$ and W is not paracompact. Thus any paracompact manifold modeled on open subsets of X will have open submanifolds which are not paracompact. A theory of such manifolds would be awkward at best.

We will see in Chapter 7 that Bw* manifolds do not share this property. If M is a paracompact Bw* manifold, then every open manifold of M will be shown to be paracompact.

The basic theory of differential calculus in bw* spaces makes no distinction between those bw* spaces with separable duals and those with non-separable duals. Thus the theory of Chapters 4 and 5 is presented for general bw* spaces. However, the facts about bw* spaces just mentioned, plus Remark 2.30 about manifolds of maps, suggest that the class of Bw* spaces, and not the full class of bw* spaces, is the proper category on which to develop an abstract model for function spaces.

3. DIFFERENTIAL CALCULUS

The fundamental results of the differential calculus of maps between finite-dimensional vector spaces are:

(1) the closure of C^k maps under composition, $k \in N$.

(2) the inverse function theorem.

(3) the existence of C^k flows for C^k vector fields.

(4) the existence, for each open subset U of R^n, of a smooth function $f : R^n \to R$ such that $f(x) \neq 0$ $\Longleftrightarrow$ $x \in U$ (which implies the existence of smooth partitions of unity).

The extension of differential calculus to Banach spaces has been completely successful, in that (1), (2), (3) are still true in the more general setting; and even (4) holds, at least for separable Hilbert space and certain other well-behaved separable Banach spaces (due to J. Wells, unpublished).

A theory of differential calculus for mappings between objects in a category of spaces properly larger than the category of Banach spaces must at least satisfy (1), and should reduce to the usual theory for mappings between Banach spaces. The usual elementary results such as Taylor's theorem should generalize.

However, the fundamental purpose of such a theory must be to provide generalizations of (2) and (3) which are of use in analysis. If there exist generalizations of (4), so much the better.

In this chapter, we propose a general theory of differential calculus, and define a category of spaces and maps for which we can prove a generalization of (1). The definitions have the advantage of being intuitive and easy to work with, and the category of spaces is large enough to include $C^\infty(D^n,R)$, the space of distributions on D^n, and the bw* spaces.

We will not prove all of the elementary theorems of differential calculus. Our goal in this chapter is simply to develop those properties of differential calculus which will be essential to our study of maps between bw* spaces in the next two chapters.

The material in this chapter would not have been possible in its present form had the author not read the paper [20] of J. Kijowski and W. Szczyrba. These authors were concerned exclusively with differential calculus in the category of Frechet Schwartz spaces, and in their theory $L^r(Z_1,Z_2)$ is given the topology of uniform convergence at each point of Z_1 rather than uniform convergence on bounded subsets of Z_1. However, many of the ideas in this chapter had their origins in [20]*.

It will be useful for us to first obtain a few basic results about inverse limits and multilinear maps.

Let $\mathcal{A}$ be a category, and Λ a directed set. Then Λ may be considered as a category whose objects are elements of Λ, and whose morphisms are defined as follows:

$$\mathrm{Map}(\alpha,\beta) = \begin{cases} \text{the set whose single element is the ordered pair } (\alpha,\beta), & \text{if } \alpha \le \beta . \\ \emptyset \text{ otherwise.} \end{cases}$$

If $\alpha \le \beta$ and $\beta \le \gamma$, then $\mathrm{Map}(\beta,\gamma) \times \mathrm{Map}(\alpha,\beta) \to \mathrm{Map}(\alpha,\gamma)$ is obviously defined by: $(\beta,\gamma)\circ(\alpha,\beta) = (\alpha,\gamma)$.

<u>3.1 Definition</u>. An inverse system F in $\mathcal{A}$ directed by Λ is a contravariant functor F from Λ to $\mathcal{A}$, i.e. a function which assigns an object $F(\lambda)$ to each $\lambda \in \Lambda$, and to each $\alpha \le \beta$, a morphism $F_{\alpha\beta} \in \mathrm{Map}(F(\beta),F(\alpha))$ such that, if $\alpha \le \beta \le \gamma$, $F_{\alpha\beta}\circ F_{\beta\gamma} = F_{\alpha\gamma}$.

<u>3.2 Definition.</u> If Λ is a directed set, and F an inverse system in $\mathcal{A}$ directed by Λ, then an inverse limit of F is an

*Since the completion of this research, W. Szczyrba has modified his earlier work with J. Kijowski to produce a theory more along the lines of the one presented here. Refer to [43] for comparison.

object $F(\infty)$ together with a set of morphisms $\{p_\alpha \in \mathrm{Map}(F(\alpha)) \mid \alpha \in \Lambda\}$ such that:

(1) for each pair $\alpha \leq \beta$, $F_{\alpha\beta} \circ p_\beta = p_\alpha$.

(2) if A is an object in $\mathcal{A}$, and $\{g_\alpha \in \mathrm{Map}(A, F(\alpha)) \mid a \in \Lambda\}$ a set of morphisms such that $F_{\alpha\beta} \circ g_\beta = g_\alpha$ for all $\alpha \leq \beta$, then there exists a unique $g \in \mathrm{Map}(A, F(\infty))$ such that $p_\alpha \circ g = g_\alpha$ for all $\alpha \in \Lambda$.

It is standard and immediate that, if $(\tilde{F}(\infty), \{\tilde{p}_\alpha \in \mathrm{Map}(\tilde{F}(\infty), F(\alpha)) \mid \alpha \in \Lambda\})$ is another inverse limit of F in $\mathcal{A}$, there exists a unique isomorphism $p \in \mathrm{Map}(F(\infty), \tilde{F}(\infty))$ such that $\tilde{p}_\alpha \circ p = p_\alpha$ and $\tilde{p}_\alpha = p_\alpha \circ p^{-1}$ for all $\alpha \in A$ (i.e. $\tilde{F}(\infty)$ and $F(\infty)$ are canonically isomorphic), so we may speak of THE inverse limit of the inverse system F, if one exists. If $F(\infty)$ is the object of the inverse limit of F, we will refer to $F(\infty)$ as the inverse limit, leaving understood that there is a distinguished set of morphisms associated with $F(\infty)$. We also denote $F(\infty)$ by $\varprojlim_{\lambda \in \Lambda} F(\lambda)$.

3.3 <u>Lemma</u>. Let F be an inverse system in the category of locally convex spaces. Then the inverse limit $F(\infty)$ of F exists. Furthermore, the underlying topological space of $F(\infty)$ is the inverse limit of F in the category of topological spaces, and the underlying set of $F(\infty)$ is the inverse limit of F in the category of sets.

<u>Proof</u>. Standard. □

Let Z be a locally convex space, and let Λ be the directed set of continuous semi-norms on Z. For each $\lambda \in \Lambda$, let $N_\lambda = \{z \in Z \mid \lambda(z) = 0\}$, and let Z_λ be the normed linear space whose underlying vector space is Z/N_λ, equipped with the norm induced by λ.

A natural question to ask is, whether or not Z is equal to

$\varprojlim Z_\lambda$. This question is discussed on pp. 230-232 of [21] and also on pp. 51-54 of [39]. It is easy to see (and is shown in each of these references) that Z may be considered as a densely embedded linear subspace of $\varprojlim Z_\lambda$. As a consequence, if Z is complete, then it follows that the two spaces must be equal. Also, if there exists a continuous norm on Z (i.e. a semi-norm λ with $\operatorname{Ker} \lambda = \{0\}$), or equivalently, if there exists a continuous linear injection of Z into a normed space, then it is easy to see that $Z = \varprojlim_{\lambda \in \Lambda} Z_\lambda$.

3.4 Definition. A locally convex space Z will be called projective if $Z = \varprojlim_{\lambda \in \Lambda} Z_\lambda$.

A simple example of a locally convex space which is not projective may be obtained as follows: let Z be any locally convex space on which the topology is generated by linear functionals (for instance, take a locally convex space and retopologize it with its weak topology). Then the kernel of each continuous semi-norm λ on Z has finite co-dimension, and so each Z_λ is finite-dimensional, and hence complete. Since $\varprojlim_{\lambda \in \Lambda} Z_\lambda$ is a closed subspace of the product of the Z_λ's, it follows that $\varprojlim_{\lambda \in \Lambda} Z_\lambda$ is complete. So if we choose a space Z which is not complete, then it follows that Z cannot be all of $\varprojlim_{\lambda \in \Lambda} Z_\lambda$.

Since no infinite-dimensional Banach space is complete in its weak topology (it is easy to construct Cauchy nets which do not converge), we have a multitude of examples of non-projective locally convex spaces. Or, even more simply, the weak topology on any incomplete normed space is also incomplete, since a sequence which is Cauchy in norm will also be weakly Cauchy.

Let V and Z be locally convex spaces, and let $L^r(V,Z)$ be the linear space of r-linear continuous maps $f : V^r \to Z$. We define the topology on $L^r(V,Z)$ to be generated by the following set of semi-norms: $M = \{\mu(B,\lambda) \mid B$ is a bounded subset of V, λ a semi-norm on $Z\}$, where for each $f \in L^r(V,Z)$, $\mu = \mu(B,\lambda)$ is defined by

$$\mu(f) = \sup\{\lambda(f(v_1,\ldots,v_r)) \mid v_i \in B,\ 1 \le i \le r\}.$$

Note that, for each semi-norm λ on Z, we have a natural projection $q_\lambda^r : L^r(V,Z) \to L^r(V,Z_\lambda)$. By our definitions of the semi-norms on $L^r(V,Z)$ and $L^r(V,Z_\lambda)$, it is immediate that the topology on $L^r(V,Z)$ is the coarsest topology such that q_λ^r is continuous for all $\lambda \in \Lambda$.

These remarks generalize naturally to inverse systems. Thus, if Λ is a general directed set, F an inverse system of locally convex spaces directed by Λ, and V a locally convex space, then the topology on $L^r(V,F(\infty))$ is the coarsest such that the map

$$q_\lambda^r : L^r(V,F(\infty)) \to L^r(V,F(\lambda))$$

is continuous for all $\lambda \in \Lambda$.

<u>3.5 Lemma.</u> $L^r(V,F(\infty)) = \lim\limits_{\substack{\leftarrow \\ \lambda\in\Lambda}} L^r(V,F(\lambda))$.

<u>Proof</u>. Let Y be a locally convex space, $\{f_\lambda : Y \to L^r(V,F(\lambda)) \mid \lambda \in \Lambda\}$ a collection of continuous linear maps such that for each pair $\alpha \le \beta$, $q_{\alpha\beta}^r \circ f_\beta = f_\alpha$. We must show there exists a unique continuous linear map $f : Y \to L^r(Y,F(\infty))$ such that $q_\alpha^r \circ f = f_\alpha$ for all $a \in \Lambda$. For the moment, fix $y \in Y$, and consider $\{f_\alpha(y) \mid a \in \Lambda\}$. For each $\alpha \in \Lambda$, $f_\alpha(y) \in C^0(V^r,F(\alpha))$. Since $F(\infty)$ is the inverse limit of F in the category of topological spaces, there exists a unique $s_y \in C^0(V^r,F(\infty))$ such that $p_\alpha \circ s_y = f_\alpha(y)$ for all $a \in \Lambda$. Thus our map f is unique, if it exists, and $f(y) = s_y$. We must show that $s_y \in L^r(V,F(\infty))$. To see

this, consider the r-tuple $(v_1,\dots,v_{i-1}, av_i+w_i,v_{i+1},\dots,v_r) \in V^r$. Note that, for each $\alpha \in \Lambda$,

$$\begin{aligned}
&p_\alpha(s_y(v_i,\dots,v_{i-1},av_i+w_i,v_{i+1},\dots,v_r)) \\
&\qquad = f_\alpha(v_1,\dots,v_{i-1},av_i+w_i,v_{i+1},\dots,v_r) \\
&\qquad = af_\alpha(v_1,\dots,v_r) + f_\alpha(v_1,\dots,w_i,\dots,v_r) \\
&\qquad = ap_\alpha(s_y(v_1,\dots,v_r)) + p_\alpha(s_y(v_1,\dots,w_i,\dots,v_r)) \\
&\qquad = p_\alpha(as_y(v_1,\dots,v_r) + s_y(v_1,\dots,w_i,\dots,v_r)).
\end{aligned}$$

Since $F(\infty) = \varprojlim_{\lambda\in\Lambda} F(\lambda)$, this implies that

$$s_y(v_1,\dots,av_i+w_i,\dots,v_r) = as_y(v_1,\dots,v_r) + s_y(v_1,\dots,w_i,\dots,v_r)$$

i.e. $s_y \in L^r(V,F(\infty))$. Similarly, it is easily seen that $s_{(ay_1+y_2)} = as_{y_1} + s_{y_2}$, so that f is linear. And since the topology on $L^r(V,F(\infty))$ is the coarsest such that q^r_λ is continuous for all $\lambda \in \Lambda$, the continuity of $q^r_\lambda \circ f = f_\lambda$ for all $\lambda \in \Lambda$ implies the continuity of f. □

Let $r \in N$. We will now compare the spaces $L(V,L^r(V,Z))$ and $L^{r+1}(V,Z)$. Let $f \in L^{r+1}(V,Z)$: fix $v \in V$, and consider the r-linear map $f_v : V^r \to Z$ defined by $f_v(w) = f(v,w)$ for all $w \in V^r$. Obviously f_v is continuous, and $f_{(v_1+v_2)} = f_{v_1} + f_{v_2}$ for all $v_1,v_2 \in V$.

Define $\tilde{f} : V \to L^r(V,Z)$ by $\tilde{f}(v) = f_v$ for all $v \in V$. To see that $\tilde{f} \in L(V,L^r(V,Z))$, we must show that $\tilde{f}$ is continuous. Let $\mu = \mu(B,\lambda)$ be one of the generating family M of semi-norms for the topology on $L^r(V,Z)$. Since $f : V^{r+1} \to Z$ is continuous, there exists a continuous semi-norm ν on V such that $\lambda(f(v_0,v_1,\dots,v_r)) \le \nu(v_0)\dots\nu(v_r)$ for all $v_0,\dots,v_r \in V$. We may assume for simplicity that $\nu(v) \le 1$ for all $v \in B$.

Then $\mu(f_v) = \sup\{\lambda(f(v,v_1,\dots,v_r)) \mid v_i \in B,\ 1 \le i \le r\}$ $\le \nu(v)$, $\Longrightarrow \tilde{f}$ is continuous.

Thus we have a canonical linear injection $L^{r+1}(V,Z) \to L(V,L^r(V,Z))$. We will next show that this injection is a topological embedding. For note that a generating set of semi-norms on $L(V,L^r(V,Z))$ is given by $\{\nu(B_1,\mu) \mid B_1$ is a bounded subset of V and $\mu \in M\}$. Restricting this set of semi-norms to $L^{r+1}(V,Z)$, we get $\{\nu(B_1 \times B,\lambda) \mid B_1$ and B are bounded in V, λ a semi-norm on $Z\}$, where $\nu(B_1 \times B,\lambda)(f)$ $=\sup\{\lambda(f(v,w)) \mid v \in B_1, w \in B^r\}$ for each $f \in L^{r+1}(V,Z)$. But this set of semi-norms is equivalent to the original set of semi-norms on $L^{r+1}(V,Z)$. Thus we have proved:

3.6 Lemma. Let V and Z be locally convex spaces. Then $L^{r+1}(V,Z) \subset L(V,L^r(V,Z))$.

3.7 Lemma. Let V and Z be locally convex spaces, and assume that V is compactly generated. If Z is sequentially complete (resp. quasi-complete, complete), then $L(V,Z)$ is sequentially complete (resp. quasi-complete, complete).

Proof. Standard. □

Let V and Z be locally convex spaces, U an open subset of V with $0 \in U$, and $h : U \to Z$ a mapping of sets:

3.8 Definition. h is tangent to 0 at 0 (or, h has 1-jet zero at 0) if:

(1) $h(0) = 0$.

(2) for each semi-norm $\| \|_2$ on Z, there exists a semi-norm $\| \|_1$ on V such that:
for each $\varepsilon > 0$, there exists a neighborhood U_ε of 0 with $U_\varepsilon \subset U$, such that $y \in U_\varepsilon \Longrightarrow \|h(y)\|_2 \le \varepsilon\|y\|_1$.

Notation. If h is tangent to 0 at 0, we say that h is $o(\|y\|)$.

3.9 Remarks. (a) h tangent to 0 at $0 \Longrightarrow h$ is continuous at 0.

(b) Assume h is tangent to 0 at 0, $i : V_1 \to V$ is continuous and linear, and $\ell : Z \to Z_1$ is continuous and linear. Then $\ell \circ h \circ i$ is tangent to 0 at 0.

(c) If V and Z are normed spaces, the above definition is equivalent to the usual condition $\|h(y)\| < \|y\| \cdot \alpha(\|y\|)$, with $\alpha(r) \to 0$ as $r \to 0$.

(d) h is tangent to 0 at $0 \iff p_\lambda \circ h$ is tangent to 0 at 0 for each semi-norm λ on Z, where $p_\lambda : Z \to Z_\lambda$.

3.10 Definition. Let U be an open subset of V, $f : U \to Z$ a mapping of sets, and $x \in U$. We say that f is (Fréchet) differentiable at x if there exists $\ell \in L(V,Z)$ such that r_x is tangent to 0 at 0, where r_x is the map defined by $r_x(y) = f(x+y) - f(x) - \ell(y)$.

3.11 Remarks. (a) If f is differentiable at x, then the map ℓ is uniquely determined, and is called the derivative of f at x (denoted by Df_x).

(b) If f is differentiable at x, then f is continuous at x.

(c) If V and Z are normed spaces, then the above definition is equivalent to the usual definition of differentiability in normed spaces.

(d) If $\ell \in L(Z,Z_1)$, then $\ell \circ f$ is differentiable at x, and $D(\ell \circ f)_x = \ell \circ Df_x$.

(e) If $f \in L(V,Z)$, then f is differentiable at x for all $x \in V$, and for each $x \in V$, $Df_x = f$.

3.12 Lemma. Let Z be a projective locally convex space. Then f is differentiable at $x \Leftrightarrow p_\lambda \circ f$ is differentiable at x for every semi-norm λ on Z.

Proof. If f is differentiable at x, then $(p_\lambda \circ f)$ is differentiable at x for all $\lambda \in \Lambda$ by 3.11 (d).

Conversely, assume $(p_\lambda \circ f)$ is differentiable at x for all $\lambda \in \Lambda$, and let $\ell_\lambda = D(p_\lambda \circ f)_x$ for all $\lambda \in \Lambda$. By 3.11 (d),

$q_{\alpha\beta} \circ \ell_\beta = \ell_\alpha$ for each pair of semi-norms α, β such that $\alpha \leq \beta$. Since $L(V,Z) = \varprojlim_{\lambda \in \Lambda} L(V,Z_\lambda)$, there exists a unique $\ell \in L(V,Z)$ such that $q_\lambda \circ \ell = \ell_\lambda$ for all $\lambda \in \Lambda$. Consider the function r_x defined by $r_x(y) = f(x+y) - f(y) - \ell(y)$. $(p_\lambda \circ r_x)(y) = (p_\lambda \circ f)(x+y) - (p_\lambda \circ f)(x) - \ell_\lambda(y)$. Since $p_\lambda \circ f$ is assumed to be differentiable at x, $p_\lambda \circ r_x$ is tangent to 0 at 0 for all $\lambda \in \Lambda$. By 3.9 (d), this implies that r_x is tangent to 0 at 0, so f is differentiable at x. □

<u>3.13 Lemma</u>. Let V_1, V_2, Z be locally convex spaces, U_i an open subset of V_i, $i = 1,2$, $f : U_1 \to U_2$ and $g : U_2 \to Z$ mappings of sets, and $x \in U_1$. If f is differentiable at x, and g is differentiable at $f(x)$, then $(g \circ f)$ is differentiable at x, and $D(g \circ f)_x = Dg_{f(x)} \circ Df_x$.

<u>Proof</u>. Standard. □

Let V, Z be locally convex spaces, U an open subset of V, $f : U \to Z$, and assume that f is differentiable at each point of U. Let $Df : U \to L(V,Z)$ be the map which assigns to each $x \in U$, the derivative of f at x.

<u>3.14 Definition</u>. $f : U \to Z$ is C^1 if f is differentiable at each point in U, and if $Df : U \to L(V,Z)$ is continuous. f is C^k, $k > 1$, if Df is C^{k-1}. f is C^∞ if f is C^k for all $k \in N$.

Note that, if $f \in L(V,Z)$, then f is a C^∞ map from V to Z.

<u>3.15 Lemma</u>. Let $\ell_1 \in L(Z,Z_1)$, $\ell_2 \in L(V_1,V)$, and assume that f is C^1. Then $\ell_1 \circ f \circ \ell_2 : \ell_2^{-1}(U) \to Z_1$ is C^1, and $D(\ell_1 \circ f \circ \ell_2)_x = \ell_1 \circ Df_{\ell_2(x)} \circ \ell_2$ for all $x \in \ell_2^{-1}(U)$.

<u>3.16 Lemma</u>. Assume that Z is projective. Then f is $C^1 \Leftrightarrow p_\lambda \circ f$ is C^1 for all semi-norms λ on Z.

Proof. If f is C^1, $(p_\lambda \circ f)$ is C^1 for all $\lambda \in \Lambda$ by the above lemma. Conversely, if $(p_\lambda \circ f)$ is C^1 for all $\lambda \in \Lambda$, then $(Df)_x$ exists for all $x \in U$ by 3.12, and $p_\lambda \circ Df = D(p_\lambda \circ f) : U \to L(V,Z_\lambda)$ is a continuous map. Since $L(V,Z) = \varprojlim_{\lambda\in\Lambda} L(V,Z_\lambda)$ in the category of topological spaces, the continuity of $p_\lambda \circ Df$ for all $\lambda \in \Lambda$ implies that Df is continuous. □

Let Z be a locally convex space, and let $i : Z \to \tilde{Z}$ be the embedding of Z in its sequential completion. Let $\alpha : [0,1]$ be a continuous curve in Z. Since Z might not be sequentially complete, $\int_0^1 \alpha(t)\,dt$ may not exist. However, let $\tilde{\alpha} = i \circ \alpha$. Since $\tilde{Z}$ is sequentially complete, $\int_0^1 \tilde{\alpha}(t)\,dt$ does exist. Suppose $\int_0^1 \tilde{\alpha}(t)\,dt \in Z$. Then, since the map i is an embedding, $\int_0^1 \alpha(t)\,dt$ exists, and $i(\int_0^1 \alpha(t)\,dt) = \int_0^1 \tilde{\alpha}(t)\,dt$. Conversely, if $\int_0^1 \alpha(t)\,dt$ exists, then $\int_0^1 \tilde{\alpha}(t)\,dt \in Z$.

3.17 Lemma. Assume $f : U \to Z$ is C^1, and that U is convex. Then $\int_0^1 Df(x+ty)(y)\,dt$ exists for all $x, x+y \in U$, and

$$f(x+y) - f(x) = \int_0^1 Df(x+ty)(y)\,dt$$

Proof. Embed Z in its completion $\tilde{Z}$, and let $\tilde{f} : U \to \tilde{Z}$ be the induced C^1 map. Note that $\tilde{Df} = i \circ Df$. If we prove the result for $\tilde{f}$, then $\int_0^1 \tilde{Df}(x+ty)(y)\,dt = f(x+y) - f(x) \in Z$, so the integral in the statement of the lemma exists, and the desired equality follows from the corresponding equality for $\tilde{f}$. Thus it suffices to prove the lemma in the case when Z is complete. But for this case, use Lemma 3.15 to reduce the problem to the case $V = Z = \mathbb{R}$, where it is the Fundamental Theorem of Integral Calculus.□

Let E be a normed space, and consider the space $L^r(V,E)$. For each semi-norm ν on V, define a subadditive function $\tilde{\nu} : L^r(V,E) \to \mathbb{R}^+ \cup \{\infty\}$ by $\tilde{\nu}(f) = \sup\{\|f(x_1,\ldots,x_r)\| \,|\, \nu(x_i) \le 1,\ 1 \le i \le r\}$. Note that $\tilde{\nu}(f) = 0 \iff f = 0$.

3.18 Definition. $L^r_\nu(V,E)$ is the "normed" linear space $\{f \in L^r(V,E) \mid \tilde{\nu}(f) < \infty\}$, with $\tilde{\nu}$ as the norm.

3.19 Remarks. (a) The inclusion $L^r_\nu(V,E) \to L^r(V,E)$ is continuous.

(b) For each $f \in L^r(V,E)$, there exists $\nu = \nu(f)$ such that $f \in L^r_\nu(V,E)$.

Note that there exists a canonical linear map $L^r(V_\nu,E) \to L^r_\nu(V,E)$, defined by $\tilde{f}(v_1,\ldots,v_r)$ $=f(p_\nu v_1,\ldots,p_\nu v_r)$ for each $f \in L^r(V_\nu,E)$, and it is obvious that $\|\tilde{f}\| = \|f\|$.

3.20 Lemma. The canonical map $L^r(V_\nu,E) \to L^r_\nu(V,E)$ is an isomorphism of normed spaces.

Proof. By the above remarks, it suffices to prove the map surjective. Let $f \in L^r_\nu(V,E)$, and consider the r-tuple $(v_1,\ldots,v_r) \in V^r$. If there exists $i \in \{1,\ldots,r\}$ such that $\nu(v_i) = 0$, then $f(v_1,\ldots,v_r) = 0$. So if $(x_1,\ldots,x_r)$ and $(y_1,\ldots,y_r)$ are two elements of V^r such that $p_\nu(x_i) = p_\nu(y_i)$ for all $1 \le i \le r$, then

$$f(x_1,\ldots,x_r) = f(x_1-y_1,x_2,\ldots,x_r) + f(y_1,x_2,\ldots,x_r)$$
$$=\Sigma^r_{i=1} f(y_1,\ldots,y_{i-1},x_i-y_1,x_{i+1},\ldots,x_r)$$
$$+f(y_1,\ldots,y_r) = f(y_1,\ldots,y_r).$$

Thus f induces a map from $(V_\nu)^r$ to Z, which is obviously r-linear and continuous. □

3.21 Definition. A D-space V is a locally convex space with the following property: if U is a neighborhood of zero in V, E a normed space, $r \in N$, and $f : U \to L^r(V,E)$ a continuous map, then there exists a neighborhood W of zero, $W \subseteq U$, and a semi-norm ν on V, such that $f(W) \subset L^r_\nu(V,E)$ and such that the map $f : W \to L^r_\nu(V,E)$ is continuous.

Note that the basic property for D-spaces is equivalent to the following property: if E is a normed space, U an open subset of V, $r \in N$, and $f : U \to L^r(V,E)$ a continuous map, then, for each $x \in U$, there exists a neighborhood U_x of x and a semi-norm $\nu = \nu(x)$ on V, such that $f(U_x) \subset L^r_\nu(V,E)$ and such that the map $f : U_x \to L^r_\nu(V,E)$ is continuous.

The obvious examples of D-spaces are the normed spaces. Although there exist highly non-trivial examples of D-spaces, we shall defer discussion of these until we have developed the differential calculus of maps between D-spaces.

3.22 Lemma. Let V be a D-space, Z any locally convex space, $r \in N$. Then $L(V,L(V,\ldots,L(V,Z))\ldots) = L^r(V,Z)$.

Proof. We proceed by induction. The result is a tautology for $r = 1$, so let $r \geq 1$ and assume the result to be true for r. Then $\underbrace{L(V,L(V,\ldots,L(V,Z))\ldots)}_{(r+1)\text{ times}} = L(V,L^r(V,Z))$, so we need only show that $L(V,L^r(V,Z)) = L^{r+1}(V,Z)$. By Lemma 3.6, it is sufficient to show that $f \in L(V,L^r(V,Z)) \Longrightarrow f \in L^{r+1}(V,Z)$, and by Lemma 3.5 it suffices to verify that $q^r_\lambda \circ f \in L^{r+1}(V,Z_\lambda)$ for all $\lambda \in \Lambda$. Let $f_\lambda = q^r_\lambda \circ f$. Obviously $f_\lambda \in L(V,L^r(V,Z_\lambda))$, so by the definition of D-space, there exists a neighborhood U of zero and a semi-norm ν on V such that $f_\lambda(U) \subset L^r_\nu(V,Z_\lambda)$ and such that the map $f_\lambda : U \to L^r_\nu(V,Z_\lambda)$ is continuous. Since f_λ is linear, we may assume $U = V$. Since $L^r_\nu(V,Z_\lambda)$ is a normed space, there exists a semi-norm μ on V such that $\|f_\lambda(x)\|_{\tilde{\nu}} \leq 1$ for all $x \in V$ with $\mu(x) \leq 1$, and we may assume $\mu \geq \nu$. Thus $\|f_\lambda(v_0,\ldots,v_r)\| \leq 1$ for all $(r+1)$-tuples $(v_0,\ldots,v_r) \in V^{r+1}$ such that $\mu(v_i) \leq 1$, $0 \leq i \leq r$, which implies $f_\lambda \in L^{r+1}(V,Z_\lambda)$. □

3.23 Proposition. Let V be a compactly generated D-space, Z a locally convex space. If Z is sequentially complete (resp. quasi-complete, complete), then $L^r(V,Z)$ is sequentially complete (resp. quasi-complete, complete) for all $r \in N$.

Proof. We already have the result for $r = 1$ by Lemma 3.7. Let $r \geq 1$, and assume the result is true for r. Since $L^{r+1}(V,Z) = L(V,L^r(V,Z))$, the result follows for $(r+1)$ by another application of 3.7. □

Our first major result for D-spaces will be that, if $f : U \to Z$ is a map, and U an open subset of V, then f is $C^k \iff p_\lambda \circ f$ is C^k for all semi-norms λ on Z. This is essentially a consequence of 3.22, which implies that $D^r f(U) \subset L^r(V,Z)$ for all $r \leq k$. But first we will need to make a couple of trivial observations, which are natural generalizations of 3.9 (d) and 3.16.

3.24 Lemma. Assume Λ is a directed set, F an inverse system of locally convex spaces directed by Λ. Let V be a locally convex space, U a neighborhood of 0 in V, and $h : U \to F(\infty)$. Then h is tangent to 0 at $0 \iff p_\lambda \circ h$ is tangent to 0 at 0 for all $\lambda \in \Lambda$.

Proof. Assume $p_\lambda \circ h$ is tangent to 0 at 0 for all $\lambda \in \Lambda$, and let μ be a semi-norm on $F(\infty)$. Since the topology on $F(\infty)$ is the coarsest such that the p_λ are all continuous, there exist $\lambda_1,\ldots,\lambda_r$ and semi-norms μ_i on $F(\lambda_i)$ such that $\mu(x) \leq \Sigma_{i=1}^r \mu_i(p_{\lambda_i} x)$ for all $x \in F(\infty)$. For each semi-norm μ_i, choose a semi-norm ν_i on V as in Definition 3.8, and let $\nu = \nu_1 + \ldots + \nu_r$. For each $\varepsilon > 0$, choose a neighborhood $U_{i\varepsilon}$ of 0, such that $y \in U_{i\varepsilon} \implies \mu_i(p_{\lambda_i} \circ h(y)) \leq \varepsilon \cdot \nu_i(y)$, and let $U_\varepsilon = \bigcap_{i=1}^{r} U_{i\varepsilon}$. Then $\mu(h(y)) \leq \varepsilon \cdot \nu(y)$ for all $y \in U_\varepsilon$, $\implies$ h is tangent to 0 at 0. □

3.25 Lemma. Assume Λ is a directed set, F an inverse system of locally convex spaces directed by Λ. Let V be a locally convex space, U an open subset of V, and $f: U \to F(\infty)$. Then f is $C^1 \iff p_\lambda \circ f$ is C^1 for all $\lambda \in \Lambda$.

Proof. Using 3.24 in place of 3.9 (d), the proof is otherwise identical to the proof of 3.16. □

3.26 Theorem. Assume Λ is a directed set, F an inverse system of locally convex spaces, V a D-space, U an open subset of V, $f : U \to F(\infty)$, and $k \in N$. Then f is $C^k \iff p_\lambda \circ f$ is C^k for all $\lambda \in \Lambda$.

Proof. We already have the result if $k = 1$, so let $k > 1$. Assume f is C^k. Then $D^r f : U \to L^r(V,F(\infty))$ is C^1 for all $r < k$, and $D(D^r f) = D^{r+1} f : U \to L^{r+1}(V,F(\infty))$. Since $q_\lambda^r: L^r(V,F(\infty)) \to L^r(V,F(\lambda))$ is linear, 3.15 implies that $q_\lambda^r \circ D^r f$ is C^1 if $r < k$, and $D(q_\lambda^r \circ D^r f) = q_\lambda^{r+1} \circ D^{r+1} f$. Thus $p_\lambda \circ f$ is C^k, and $D^r(p_\lambda \circ f) = q_\lambda^r \circ (D^r f)$ for all $r \leq k$.

Conversely, assume $p_\lambda \circ f$ is C^k for all $\lambda \in \Lambda$. For each $r \leq k$, consider the system of maps $D^r(p_\lambda \circ f) : U \to L^r(V,F(\lambda))$. Applying the first part of this proof, we conclude that $q_{\alpha\beta}^r(D^r(p_\beta \circ f)) = D^r(p_\alpha \circ f)$ for each pair $\alpha,\beta \in \Lambda$ with $\alpha \leq \beta$. By Lemma 3.5, there exists a unique continuous map $A_r: U \to L^r(V,F(\infty))$ such that $q_\lambda^r \circ A_r = D^r(p_\lambda \circ f)$ for all $\lambda \in \Lambda$. By Lemma 3.25, A_r is C^1 if $r < k$. Now, for $r < k$, $q_\lambda^{r+1} \circ (DA_r) = D(q_\lambda^r \circ A_r) = D(D^r(p_\lambda \circ f)) = D^{r+1}(p_\lambda \circ f) = q_\lambda^{r+1} \circ A_{r+1}$, $\implies DA_r = A_{r+1}$, $\implies f = A_0$ is C^k. □

3.27 Corollary. Let V be a D-space, Z a projective locally convex space, U an open subset of V, $f : U \to Z$, and $k \in N$. Then f is $C^k \iff p_\lambda \circ f$ is C^k for all semi-norms λ on Z.

Consider the situation of V a D-space, E a normed space, U open in V, and $f : U \to L(V,E)$ a C^k map, $k \geq 1$. We know, by definition, that for each $x \in V$, there exists a semi-norm ν on V and a neighborhood U_x of x such that $f(U_x) \subset L_\nu(V,E)$ and $f: U_x \to L_\nu(V,E)$ is continuous. Our next goal is to show that U_x and ν can be chosen so that $f : U \to L_\nu(V,E)$ is C^k. Once we have this result, the closure of C^k functions under composition will be immediate. However, we first must establish a few technical results. The first is a converse to 3.17.

3.28 Lemma. Let V be a D-space, Z a complete locally convex space, U a convex open subset of V, and $f : U \to Z$ a mapping of sets. Assume there exists a continuous map $g : U \to L(V,Z)$ such that

$$f(x+y) - f(x) = \int_0^1 g(x+ty)(y)\,dt \quad \text{for all } x,\ x+y \in U.$$

Then f is C^1, and $Df = g$.

Proof. It suffices to show that, for each $x \in U$, and semi-norm λ on Z, $Df_\lambda(x)$ exists, and $Df_\lambda(x) = g_\lambda(x)$, where $f_\lambda = p_\lambda \circ f$, $g_\lambda = q_\lambda \circ g : U \to L(V,Z_\lambda)$.

Obviously $f_\lambda(x+y) - f_\lambda(x) = \int_0^1 g_\lambda(x+ty)(y)\,dt$. Choose a neighborhood U_x of x and a semi-norm ν on X such that $g_\lambda(U_x) \subset L_\nu(V,Z_\lambda)$ and $g_\lambda : U_x \to L_\nu(V,Z_\lambda)$ is continuous, and choose a convex neighborhood U_ε of x such that $\|g_\lambda(x+y) - g_\lambda(x)\|_{\tilde{\nu}} \leq \varepsilon$ for all $x+y \in U_\varepsilon$. Then

$$\|f_\lambda(x+y) - f_\lambda(x) - g_\lambda(x)(y)\| \leq \int_0^1 \|g_\lambda(x+ty) - g_\lambda(x)\|_{\tilde{\nu}} \cdot \nu(y)\,dt \leq \varepsilon \cdot \nu(y),$$

$\Longrightarrow Df_\lambda(x)$ exists and $Df_\lambda(x) = g_\lambda(x)$. □

3.29 Lemma. Let V be a D-space, Z and Z_1 locally convex spaces, Z complete. Let U be a convex open subset of V, $f : U \to Z$ a mapping of sets, $g : U \to L(V,Z)$ a continuous map. Suppose there exists $\ell \in L(Z,Z_1)$ such that:

(1) ℓ is injective

(2) $\ell \circ f$ is C^1, and $D(\ell \circ f) = \ell \circ g$

Then f is C^1, and $Df = g$.

Proof. Since $\ell \circ f$ is C^1, and $D(\ell \circ f) = \ell \circ g$,

$$\begin{aligned} \ell(f(x+y) - f(x)) &= \ell \circ f(x+y) - \ell \circ f(x) \\ &= \int_0^1 (\ell \circ g)(x+ty)(y)\,dt \\ &= \ell\left(\int_0^1 g(x+ty)(y)\,dt\right) \quad \text{for all } x,\ x+y \in U. \end{aligned}$$

Since ℓ is injective, $f(x+y) - f(x) = \int_0^1 g(x+ty)(y)dt$. Thus by 3.28, f is C^1 and $Df = g$. □

3.30 Lemma. Let V be a D-space, Z_1 and Z_2 locally convex spaces, $\ell: Z_1 \to Z_2$ a linear topological embedding, U an open subset of V, and $f : U \to Z_1$. Assume that $\ell \circ f$ is C^1, and also that there exists a continuous map $g : U \to L(V,Z_1)$ such that $D(\ell \circ f) = \ell \circ g$.
Then f is C^1, and $Df = g$.

Proof. It is sufficient to show that, for each $x \in U$, $Df(x)$ exists and is equal to $g(x)$.

For each $x \in U$, consider the map γ_x defined by $\gamma_x(y) = f(x+y) - f(x) - g(x)(y)$. $\ell \circ \gamma_x(y) = \ell \circ f(x+y) - \ell \circ f(x) - D(\ell \circ f)(x)(y)$, $\Rightarrow$ $\ell \circ \gamma_x$ is tangent to 0 at 0. Since ℓ is a linear topological embedding, γ_x is tangent to 0 at 0, $\Rightarrow$ $Df(x)$ exists and equals $g(x)$. □

3.31 Lemma. Let V be a D-space, Z and Z_1 locally convex spaces, $\ell \in L(Z,Z_1)$, $k \in N$, U open in V, and $f : U \to Z$ a C^k map. Then $\ell \circ f$ is C^k, and $D^i(\ell \circ f) = \ell \circ D^i f$ for each $i \leq i < k$.

Proof. We already know this for $k = 1$. So assume $k > 1$. For each $i < k$, $D^i f : U \to L^i(V,Z)$ is a C^1 map. Since ℓ induces a continuous linear map from $L^i(V,Z)$ to $L^i(V,Z_1)$, $\ell \circ D^i f$ is a C^1 map from U to $L^i(V,Z_1)$, and $D(\ell \circ D^i f) = \ell \circ D(D^i f) = \ell \circ D^{i+1} f$. Thus $\ell \circ f$ is C^k, and $D^i(\ell \circ f) = \ell \circ D^i f$ for each $1 \leq i < k$. □

3.32 Theorem. Let V be a D-space, E a normed space, U an open subset of V, $k \in N \cup \{0\}$, $r \in N$, and $f : U \to L^r(V,E)$ a C^k map. Then, for each $x \in U$, there exists a semi-norm ν on V, and a neighborhood U_x of x, such that $f(U_x) \subset L^r_\nu(V,E)$, and such that the map $f : U_x \to L^r_\nu(V,E)$ is C^k.

Proof. We will proceed by induction on k. For $k = 0$, the theorem reduces to the statement that V is a D-space, and so it is trivially true in this case.

So let $k > 0$, and assume that the theorem is true for $k - 1$. Let $x \in U$. Since V is a D-space, we know there exists a semi-norm ν on V and a neighborhood U_0 of x such that $f(U_0) \subset L_\mu^r(V,E)$ and such that $f : U_0 \to L_\mu^r(V,E)$ is continuous.

By our inductive assumption, there exists a semi-norm η on V and a neighborhood U_1 of x such that $Df(U_1) \subset L_\eta^{r+1}(V,E)$, and such that $Df:U_1 \to L_\eta^{r+1}(V,E)$ is C^{k-1}.

Let $U_x = U_0 \cap U_1$, and $\nu = \mu+\eta$. Since the linear injections $L_\mu^r(V,E) \to L_\nu^r(V,E)$ and $L_\eta^r(V,E) \to L_\nu^r(V,E)$ are continuous, $f:U_x \to L_\nu^r(V,E)$ is continuous, and $Df:U_x \to L_\nu^{r+1}(V,E)$ is C^{k-1}.

Embed E in its completion E_1, which induces embeddings $L_\nu^r(V,E) \subset L_\nu^r(V,E_1)$. Note that $L_\nu^r(V,E_1) = L^r(V_\nu,E_1)$ is complete, and that the induced map $Df:U_x \to L_\nu^{r+1}(V,E_1) \to L(V,L_\nu^r(V,E_1))$ is continuous. Also, note that $f:U \to E_1$ is C^k.

Since the inclusion $L_\nu^r(V,E_1) \to L^r(V,E_1)$ is linear and injective, we may apply 3.29 to conclude that $f:U_x \to L_\nu^r(V,E_1)$ is C^1. But $Df:U_x \to L_\nu^{r+1}(V,E) \approx L_\nu(V,L_\nu^r(V,E)) \to L(V,L_\nu^r(V,E)) \to L(V,L_\nu^r(V,E_1))$ factors continuously through $L(V,L_\nu^r(V,E))$, so we may apply 3.30 to conclude that f is a C^1 map from U_x to $L_\nu^r(V,E)$.

Now, since $Df:U_x \to L_\nu^{r+1}(V,E)$ is C^{k-1}, it follows from 3.31 that $Df:U_x \to L(V,L_\nu^r(V,E))$ is also C^{k-1}. Thus $f:U_x \to L_\nu^r(V,E)$ is C^k. □

<u>3.33 Theorem.</u> Let V and Y be D-spaces, E a normed space, U_1 an open subset of V, U_2 an open subset of Y. Let $k \in \mathbb{N} \cup \{0\}$, and assume $f : U_1 \to U_2$ and $g : U_2 \to E$ are C^k maps. Then $g \circ f$ is C^k.

<u>Proof</u>. We proceed by induction. The result is point-set topology for $k = 0$, so let $k \geq 1$, and assume the theorem proved for the case $(k-1)$.

By Lemma 3.13, $D(g \circ f)(x)$ exists for each $x \in U_1$, and $D(g \circ f)(x) = Dg(f(x)) \cdot Df(x)$. Let $y = f(x)$. By the above theorem, there exists a semi-norm μ on Y and a neighborhood U_y of y such that $(Dg)(U_y) \subset L_\mu(Y,E) = L(Y_\mu,E)$ and such that

$Dg: U_y \to L_\mu(Y,E)$ is C^{k-1}.

Consider $D(p_\mu \circ f) = q_\mu \circ (Df) : U_1 \xrightarrow{C^{k-1}} L(V,Y_\mu)$. Another application of 3.32 yields the existence of a semi-norm ν on V and a neighborhood U_x of x, such that $(q_\mu \circ Df)(U_x) \subset L_\nu(V,Y_\mu) = L(V_\nu,Y_\mu)$ and such that the map $q_\mu \circ Df : U_x \to L(V_\nu,Y_\mu)$ is C^{k-1}. We will assume that $U_x \subset g^{-1}(U_y)$.

Define the bilinear map

$$A : L(Y_\mu,E) \times L(V_\nu,Y_\mu) \to L(V_\nu,E).$$
$$(g,f) \longrightarrow g\circ f$$

Note that A is a continuous bilinear map between normed spaces, which implies that A is C^∞. Also, the domain of A is a D-space.

We now have the following factorization of $D(g\circ f)$:

$$U_x \xrightarrow{((Dg)\circ f,\, q_\mu \circ Df)} L(Y_\mu,E) \times L(V_\nu,Y_\mu)$$

$$\xrightarrow{A} L(V_\nu,E) = L_\nu(V,E) \longrightarrow L(V,E)$$

Since we already have the theorem for the case $(k-1)$, $(Dg)\circ f$ is C^{k-1}, so $((Dg)\circ f,\, q_\mu \circ Df)$ is C^{k-1}, which implies that $A\circ((Dg)\circ f, q_\mu \circ Df)$ is C^{k-1}. Since $i : L_\nu(V,E) \xrightarrow{C^\infty} L(V,E)$ is linear, $D(g\circ f) = i\circ A\circ((Dg)\circ f, q_\mu\circ Df)$ is C^{k-1}, and we conclude that $g\circ f$ is C^k. □

<u>3.34 Theorem</u>. Let V and Y be D-spaces, Z a projective locally convex space, U_1 an open subset of V, U_2 an open subset of Y, and $k \in \mathbb{N} \cup \{\infty\} \cup \{0\}$. Assume that $f : U_1 \to U_2$ and $g : U_2 \to Z$ are C^k maps. Then $g\circ f$ is C^k.

<u>Proof</u>. We may assume $k \in \mathbb{N}$. Let λ be a semi-norm on Z, so $p_\lambda \circ g$ is C^k by 3.27. Thus $(p_\lambda\circ g)\circ f$ is C^k by the above theorem. Since $p_\lambda\circ(g\circ f)$ is C^k for all semi-norms λ on Z, another application of 3.27 yields that $g\circ f$ is C^k. □

<u>3.35 Proposition</u>. Let V be a D-space, Z a projective locally convex space, U an open subset of V, and $k \in \mathbb{N} \cup \{\infty\} \cup \{0\}$.

Assume $f : U \to Z$ is a C^k map, and $\alpha : U \to \mathbb{R}$ is a C^k function. Then $\alpha f : U \to Z$ defined by $(\alpha f)(x) = \alpha(x)f(x)$ is a C^k map.

Proof. It suffices to show that $p_\lambda \circ (\alpha f) = (\alpha f)_\lambda$ is C^k. Note that $p_\lambda \circ (\alpha f) = \alpha f_\lambda$. Thus we have the following factorization of $p_\lambda \circ (\alpha f)$:

$$U \xrightarrow{(\alpha, f_\lambda)} \mathbb{R} \times Z_\lambda \xrightarrow{m} Z_\lambda, \quad \text{where } m$$

is scalar multiplication. Since $\mathbb{R} \times Z_\lambda$ is a normed space, and hence a D-space, $p_\lambda \circ (\alpha f)$ is C^k by 3.34. □

Let V be a D-space, Z a locally convex space, U an open subset of V, $f : U \to L^r(V,V)$, $g : U \to L(V,Z)$. Then we will define the map $g \cdot f$ by $(g\cdot f)(x)(w) = g(x)(f(x)(w))$ for all $w \in V^r$, $x \in U$.

3.36 Proposition. Assume that $f : U \to L^r(V,V)$ and $g : U \to L(V,Z)$ are both C^k maps, $k \in \mathbb{N} \cup \{\infty\} \cup \{0\}$, and that Z is projective. Then $g\cdot f$ is C^k.

Proof. It suffices to show that $q_\lambda \circ (g\cdot f) = (q_\lambda \circ g)\cdot f$ is C^k for all semi-norms λ on Z, so we may assume that Z is a normed space. It also suffices to prove the theorem for $k \in \mathbb{N} \cup \{0\}$. So let $x \in U$. Then there exists a neighborhood U_1 of x and a semi-norm λ on V such that $g(U_1) \subset L_\lambda(V,Z) = L(V_\lambda,Z)$ and such that $g : U_1 \to L(V_\lambda,Z)$ is C^k.

Similarly, there exists a neighborhood U_2 of x and a semi-norm ν on V such that $(q_\lambda^r \circ f)(U_2) \subset L_\nu(V,V_\lambda)$ and such that the map $q_\lambda^r \circ f : U_2 \to L_\nu(V,V_\lambda)$ is C^k. Let $U_x = U_1 \cap U_2$. Then, on U_x, we have the following factorization of $g\cdot f$:

$$U_x \xrightarrow{(g, q_\lambda^r \circ f)} L(V_\lambda,Z) \times L_\nu^r(V,V_\lambda) \longrightarrow L_\nu^r(V,Z) \longrightarrow L^r(V,Z).$$

Thus $g\cdot f$ is C^k. □

3.37 Proposition. Let V be a D-space, Z a projective locally convex space, $r \in N$, and $f \in L^r(V,Z)$. Then the polynomial map $\tilde{f} : V \to Z$, defined by $\tilde{f}(x) = f(x,x,\ldots,x)$, is C^∞.

Proof. We already know this result if $r = 1$. To prove the result in general, it suffices to show that $\tilde{f}_\lambda = p_\lambda \circ \tilde{f}$ is C^∞ for all semi-norms λ on Z. But there exists a semi-norm $\nu = \nu(f)$ on V such that $f_\lambda \in L^r_\nu(V,Z_\lambda)$, so we have the following factorization of $\tilde{f}_\lambda$:

$$V \longrightarrow (V_\nu)^r \xrightarrow{f_\lambda} Z_\lambda$$

$$x \longrightarrow (p_\nu(x),\ldots,p_\nu(x)) \rightarrow f_\lambda(p_\nu(x),\ldots,p_\nu(x))$$

The first component of the factorization is linear, and the second component is a multi-linear map between normed spaces. Thus both components are C^∞, which implies that $\tilde{f}_\lambda$ is C^∞. □

3.38 Proposition. Let V be a D-space, Z a locally convex space, U an open subset of V, and $f : U \to Z$ a C^k map. Assume Z_1 is a closed linear subspace of Z such that $f(U) \subset Z_1$. Then f is a C^k map from U to Z_1.

Proof. Same as the usual proof for Banach spaces. □

This completes our discussion of differential calculus for the present. We will return to this subject in the next two chapters, but the remainder of this chapter will be devoted to the discussion of examples of D-spaces. Our examples will be exponential spaces and Schwartz spaces.

3.39 Definition. An exponential space V is a locally convex space such that V^r is compactly generated for all $r \in N$.

3.40 Remarks. Let V be a locally convex space, and Z a closed subspace of V.

(a) If V is an exponential space, then Z and V/Z are exponential spaces.

(b) If V is metrizable, then V is an exponential space.

3.41 Definition. Let Z_1, Z_2 be locally convex spaces. For each $r \in N$, we define $ev: L^r(Z_1,Z_2) \times Z_1^r \to Z_2$ to be the evaluation map, i.e. $ev(f,w) = f(w)$ for all $f \in L^r(Z_1,Z_2)$, $w \in Z_1^r$.

Our next lemma is the fundamental result about exponential spaces.

3.42 Lemma. Let V be an exponential space, Z a locally convex space, U an open subset of V, $r \in N$, and $f : U \to L^r(V,Z)$ a continuous map. Then $ev\circ(f\times I) : U\times V^r \to Z$ is continuous.

Proof. We have the following factorization of $ev\circ(f\times I)$:

$$U \times V^r \xrightarrow{f\times I} L^r(V,Z) \times V^r \longrightarrow C^0(V^r,Z) \times V^r \xrightarrow{ev} Z,$$

where $C^0(V^r,Z)$ is the space of continuous maps from V^r to Z with the compact-open topology. Since $U \times V^r$ is an open subset of V^{r+1}, it is compactly generated by 2.34. Thus it suffices to check the continuity of $ev\circ(f\times I)$ on compact subsets of $U \times V^r$. Now, the first two maps of the above factorization are continuous, and it is standard point-set topology that $ev: C^0(V^r,Z) \times V^r \to Z$ is continuous on compact subsets of $C^0(V^r,Z) \times V^r$. Thus $ev\circ(f\times I)$ is continuous.□

3.43 Definition. A Schwartz space V is a locally convex space such that, for each semi-norm λ on V, there exists a semi-norm ν on V such that $\nu \geq \lambda$ and such that the induced map $Z_\nu \to Z_\lambda$ is precompact.

We also give the two more function spaces, which we will use in the proof that an exponential Schwartz space is a D-space:

3.44 Definition. Let Z_1 and Z_2 be locally convex spaces, and let $r \in N$.

(1) $L_s^r(Z_1,Z_2)$ is the linear space of continuous r-linear maps

maps from Z_1 to Z_2 with the topology of pointwise convergence.

(3) $L^r_c(Z_1,Z_2)$ is the linear space of continuous r-linear maps from Z_1 to Z_2 with the topology of uniform convergence on precompact subsets of Z_1^r.

We will only be interested in the above spaces in the case when Z_1 and Z_2 are normed spaces. The two topologies are related by the following remark, which is an immediate consequence of the definition of a precompact set.

3.45 Remark. Let E_1 and E_2 be normed spaces, let $r \in N$, and assume that A is a bounded subset of $L^r(E_1,E_2)$. Then the subspace topology induced on A by $L^r_s(E_1,E_2)$ coincides with the subspace topology induced on A by $L^r_c(E_1,E_2)$.

3.46 Theorem. Let V be an exponential Schwartz space. Then V is a D-space.

Proof. Let E be a normed space, U an open subset of V, $r \in N$, $f : U \to L^r(V,E)$ a continuous map, and $x \in U$. We must find a neighborhood U_x of x and a semi-norm ν on V such that $f(U_x) \subset L^r_\nu(V,E)$, and such that the map $f : U_x \to L^r_\nu(V,E)$ is continuous. By 3.42 the induced map $g = ev\circ(f\times I) : U \times V^r \to E$ is continuous. Since g is continuous, and $g(x,0) = 0$, there exists a neighborhood U_x of x and a semi-norm λ on V such that $g(y,w) \le 1$ for all $y \in U_x$ and for all $w = (v_1,\ldots,v_r) \in V^r$ such that $\lambda(v_i) \le 1$ for all $1 \le i \le r$.

Thus $f(U_x) \subset L^r_\lambda(V,E) = L^r(V_\lambda,E)$. Unfortunately, f might not map U_x continuously into $L^r(V_\lambda,E)$. However, since g is continuous, the map $g_w : U_x \to E$ which is defined by

$$\begin{array}{l} U_x \xrightarrow{f} L^r(V_\lambda,E) \xrightarrow{ev_w} E \quad \text{is continuous for each } w \in V^r_\lambda, \\ y \longrightarrow f(y) \longrightarrow f(y)(w) \end{array}$$

which implies that the map $f : U_x \to L^r_s(V_\lambda,E)$ is continuous. Note that $\|f(y)\|_{\tilde{\lambda}} \le 1$ for all $y \in U_x$, so that the topology on $f(U_x)$

as a subspace of $L_s^r(V_\lambda, E)$ is the same as the topology which it inherits as a subspace of $L_c^r(V_\lambda, E)$. Thus $f : U_x \to L_c^r(V_\lambda, E)$ is continuous. Finally, choose a semi-norm ν on V such that the map $V_\nu \to V_\lambda$ is precompact, which implies that the induced map $L_c^r(V_\lambda, E) \to L^r(V_\nu, E) = L_\nu^r(V, E)$ is continuous. Then $f(U_x) \subset L_\nu^r(V, E)$, and $f : U_x \to L_\nu^r(V, E)$ is continuous. □

The definition of a Schwartz space is due to Grothendieck, and his original article [15] remains one of the best references on the subject (another very readable reference is [16]). The following stability properties of Schwartz spaces were established by Grothendieck in [15]. We omit the proof.

3.47 Lemma. (a) If V is a Schwartz space, and Z a subspace of V, then Z is a Schwartz space. If Z is closed, then V/Z is a Schwartz space.

(b) Let $F(\infty) = \lim\limits_{\substack{\to \\ n \in N}} F(n)$ be the direct limit of a sequence of Schwartz spaces. Then $F(\infty)$ is a Schwartz space.

(c) If $F(\infty) = \lim\limits_{\substack{\leftarrow \\ \lambda \in \Lambda}} F(\lambda)$ is the limit of an inverse system of Schwartz spaces, then $F(\infty)$ is a Schwartz space.

(d) If V and Z are Schwartz spaces, then $V \times Z$ is a Schwartz space.

(e) If V is a Fréchet Schwartz space, then V' is a Schwartz space.

3.48 Remark. It follows from 3.40 that a Frechet Schwartz space is a D-space. Thus $C^\infty(M)$ and $C_0^\infty(M)$, where M is any manifold (compact or non-compact), are D-spaces and so we have our first non-trivial examples of D-spaces.

Let E_1 and E_2 be Banach spaces, $f : E_1 \to E_2$ a linear map. Then, regarding each E_i as canonically embedded in its second dual E_1'', we may also regard f as a map from E_1'' to be E_2''. Under these assumptions, we have the following result:

<u>3.49 Lemma.</u> Let E_1 and E_2 be Banach spaces, $f_1: E_1 \to E_2$ a completely continuous linear map. Then $f(E_1'') \subset E_2$.

<u>Proof</u>. Consider the unit ball $B(1) \subset E_1$: since f is completely continuous, the closure of $f(B(1))$ in E_2, which we will call K, is a compact set in E_2. It is also balanced and convex. Thus $X = \varinjlim_{n\in N} K(n)$ is a Bw* space, and so we may factor f into the composition of two linear maps as follows: $E_1 \xrightarrow{\bar{f}} X \xrightarrow{i} E_2$.

Thus, as a map from E_1'' to E_2'', f may be factored into $E_1'' \xrightarrow{\bar{f}} X'' \xrightarrow{i} E_2''$. But note that the canonical embedding of X in X'' is surjective, and so $f(E_1'') = i(X'') = i(X) \subset E_2$. □

<u>3.50 Proposition</u>. Let V be a Frechet Schwartz space. Then V' is a D-space.

<u>Proof</u>. We will first show that V is compactly generated. We may write $V = \varprojlim_{n\in N} E(n)$, where each E_n is a Banach space, and the map $E(n+1) \to E(n)$ is completely continuous for all $n \in N$. Dualizing this inverse system, we obtain:
$E'(1) \to E'(2) \to E'(3) \to \dots \to V'$. I claim that V' is the direct limit of the sequence $E'(1) \to E'(2) \to \dots$ in the category of locally convex spaces. To see this, let F be a locally convex space, and let

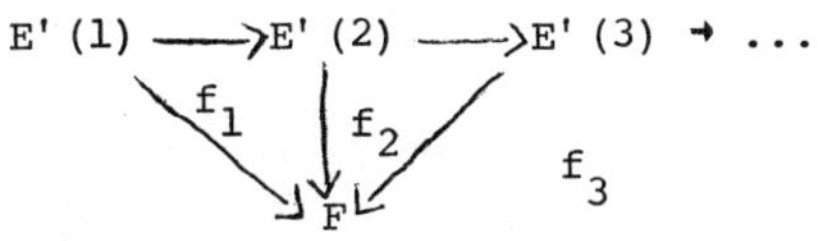

be a direct system of linear maps. Dualizing this system, we get

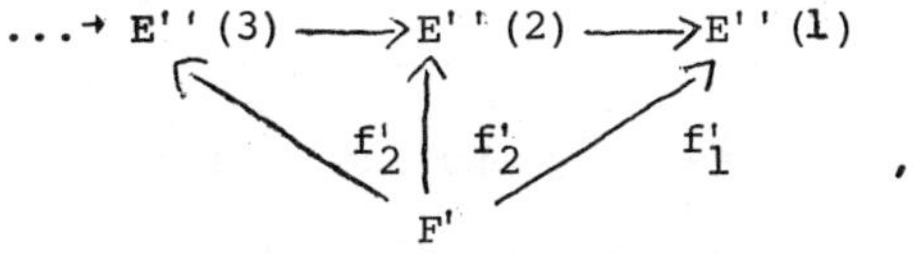

an inverse system of maps. Thus this inverse system factors through the inverse limit $\lim\limits_{\substack{\leftarrow \\ n\in N}} E''(n)$. Now, I claim that this inverse limit is V. For, since V is reflexive, and since each $E(n)$ is embedded in $E''(n)$, V is certainly canonically embedded in $\lim\limits_{\substack{\leftarrow \\ n\in N}} E''(n)$. But by Lemma 3.49, the image of each $E''(n+1)$ in the inverse system is actually contained in $E(n)$. Thus the additional vectors in the system contribute nothing to the inverse limit, and $V = \lim\limits_{\substack{\leftarrow \\ n\in N}} E''(n)$. So we may factor the inverse system $\{f'_n\}_{n\in N}$ through V.

Dualizing again, we obtain:

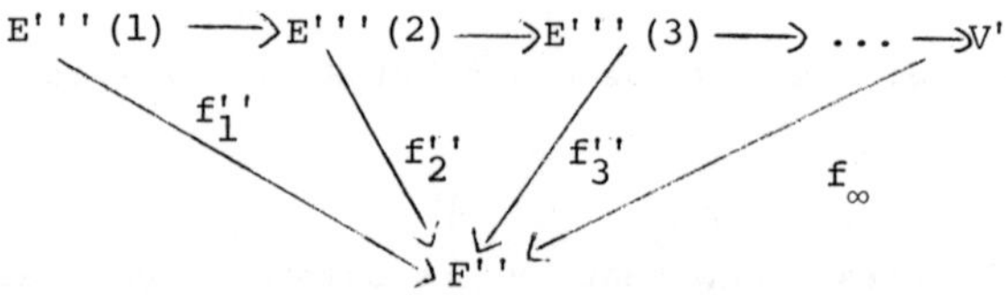

Now, each element of V' is a continuous linear functional on V, and hence extends to a functional on $E(n)$ for some $n \in N$. Thus $f_\infty(V') = \bigcup\limits_{n\in N} f''_n(E'(n)) = \bigcup\limits_{n\in N} f_n(E'(n))$, which implies that $f_\infty(V') \subset F$. Finally, since the topology which F inherits as a subspace of F'' is at least as strong as the original topology on F, f_∞ may be regarded as a continuous map from V' to F, which implies that $V' = \lim\limits_{\substack{\rightarrow \\ n\in N}} E'(n)$ in the category of locally convex spaces.

We want to be able to conclude from the above that V' is compactly generated. Unfortunately, the direct limit of a sequence of locally convex spaces in the category of locally convex spaces is not in general the same as the direct limit of the sequence in the category of topological spaces (the two direct limits can be regarded as having the same underlying set, but the topological direct limit will in general have more open subsets and will not be a locally convex space), so we will have to show directly that V' is the direct limit of $\{E'(n)\}_{n\in N}$ in the category of topological spaces. To do this, it suffices to show that the topology on the topological direct limit of $\{E'(n)\}_{n\in N}$, which has the same underlying set as V', is locally convex.

So consider the sequence $E'(1) \to E'(2) \to E'(3) \to \dots$; we may assume that the norm of each map in the sequence is less than or equal to one. Let $K(n)$ be the image in V' of the closed ball of radius n in $E'(n)$, and let $X = \lim_{\substack{\to \\ n\in N}} K(n)$. Then X has the same underlying set as V', and a few moments of abstract diagram chasing should make it clear that X is the topological direct limit of $\{E'(n)\}_{n\in N}$. Furthermore, a slight modification of the proof of Theorem 2.11 yields a proof that the topology on X is locally convex. Thus $X = V'$. Since V' is the topological direct limit of a sequence of compactly generated spaces, V' is compactly generated.

Let $r \in N$. Since $(V')^r$ is isomorphic to $(V^r)'$, and since V^r is a Frechet Schwartz space, we have that $(V')^r$ is isomorphic to the dual of a Frechet Schwartz space. Thus $(V')^r$ is compactly generated, and hence V' is an exponential space. But we already have (Lemma 3.47) that V' is a Schwartz space, so V' is an exponential Schwartz space, and hence a D-space. □

3.51 Example. Let M be a compact C^∞ manifold and let ξ be a C^∞ vector bundle over M. Then $H_{-\infty}(\xi)$, the space of distributional sections of ξ, is a D-space.

3.52 Remark. If $E(1) \to E(2) \to E(3) \to \dots$ is a direct system of Banach spaces such that each map $E(n) \to E(n+1)$ is compact, and if Z is the direct limit of this sequence in the category of locally convex spaces, then it is easy to rearrange the proof of Proposition of 3.50 to conclude: first, that Z is the topological direct limit of the sequence; and second that Z' is a Frechet Schwartz space. Since Z is the direct limit of a sequence of bornological spaces, Z is bornological. From the characterization of Z as the topological direct limit of $\{E(n)\}_{n\in N}$, it follows that closed and bounded subsets of Z are compact. Thus Z is reflexive, and hence

Z is the dual of a Frechet Schwartz space. Thus we have obtained a characterization of the duals of Frechet Schwartz spaces as exactly those locally convex spaces which are direct limits of sequences of Banach spaces $E(1) \to E(2) \to \ldots$, where each map in the sequence is compact.

3.53 Theorem. Let X be a bw* space. Then X is an exponential Schwartz space.

Proof. We know that bw* spaces are compactly generated. Since X^r is a bw* space for each $r \in N$, X is an exponential space. To see that X is a Schwartz space, recall the generating set of semi-norms given in Lemma 2.22 for the topology on X.

Each such semi-norm is of the form $\lambda(x) = \sup\{|f_i(x)| \mid i \in N\}$, where $\{f_i \mid i \in N\}$ is a sequence of elements of X' (which we may assume to be non-zero) such that $\lim_{i\to\infty} f_i = 0$. Define $r_i = \|f_i\|^{1/2}$, and let $g_i = f_i/r_i$. Then $\lim_{i\to\infty} g_i = 0$. If we define a semi-norm ν on X by $\nu(x) = \sup\{|g_i(x)| \mid i \in N\}$, it is easy to see that the map $X_\nu \to X_\lambda$ is precompact. □

We will now give an example of a large class of locally convex spaces which are not D-spaces. The demonstration that these spaces are not D-spaces is a consequence of the following well-known lemma, the proof of which we include for the sake of completeness.

3.54 Lemma. Let V be a locally convex space, and assume that the evaluation map $ev: V' \times V \to R$ is continuous. Then V is normable.

Proof. $ev^{-1}((-1,1))$ is open in $V' \times V$, so let W be an open subset of V', U an open convex balanced subset of V, such that $(0,0) \in W \times U \subset ev^{-1}((-1,1))$. Since the topology on V' has a base given by semi-norms defined by uniform convergence of bounded subsets of V, there exists a closed bounded convex balanced subset $B \subset V$ such that $\nu^{-1}([0,1)) \subset W$, where $\nu(\ell) = \sup_{x\in B}|\ell(x)|$ for each

$\ell \in V'$.

Now, assume $\ell \in V'$ such that $\nu(\ell) < 1$. Then $\ell \in W$, so $|\ell(x)| < 1$ for all $x \in U$. Since $\nu^{-1}([0,1]) = \{\ell \in V' : |\ell(x)| \leq 1 \ \forall\, x \in B\}$, it follows that $U \subset B$, and hence that U is bounded. Since V has a bounded open set, V is normable. □

3.55 Proposition. Let V be a locally convex space which is not normable. Then $V' \times V$ is not a D-space, and is not compactly generated.

Proof. Note that $L(V' \times V, R) = L(V', R) \times L(V, R) = V'' \times V'$.

Define a map f from $V' \times V$ to its dual space $V'' \times V'$ by $f((\ell, v)) = (0, \ell)$. The map f is obviously continuous.

Now, if $V' \times V$ were a D-space, the map $(V' \times V) \times (V' \times V) \to R$ defined by $(\ell_1, v_1) \times (\ell_2, v_2) \to (f(\ell_1, v_1))(\ell_2, v_2)$ would be continuous. But $(f(\ell_1, v_1))(\ell_2, v_2) = \ell_1(v_2)$, so that the continuity of this map would be equivalent to the continuity of the evaluation map $ev: V' \times V \to R$. Since V is not normable, the evaluation map is not continuous, and hence $V' \times V$ cannot be a D-space.

Furthermore, the evaluation map $ev: V' \times V \to R$ is continuous on compact subsets of $V' \times V$ (indeed, it is even continuous on subsets of $V' \times V$ of the form $V' \times C$, where C is a compact subset of V). Thus, if $V' \times V$ were compactly generated, $ev: V' \times V \to R$ would be continuous. □

3.56 Remarks. (a) Let V be an infinite-dimensional Frechet Schwartz space. Then V' is also an exponential Schwartz space, so V and V' are an example of a pair of exponential Schwartz spaces whose product, though a Schwartz space, is neither an exponential space nor a D-space. Furthermore, let $V = \lim\limits_{\substack{\leftarrow \\ n \in N}} E(n)$, so that

$V' = \lim\limits_{\substack{\rightarrow \\ n \in N}} E(n)$. Since the direct limit in the category of linear topological spaces commutes with the product operation,

$$V \times V' = \lim_{\substack{\rightarrow \\ n \in N}} (V \times E'(n))$$ in the category of linear topological spaces. Since each $V \times E'(n)$ is metrizable, and hence compactly generated, the topological direct limit of $\{V \times E'(n)\}_{n \in N}$ must itself be compactly generated. Thus we have an example of a sequence of metrizable locally convex spaces whose topological and linear topological direct limits are distinct topological spaces.

(b) Let X be an infinite-dimensional bw* space. Then X' is a Banach space, so both X and X' are D-spaces and compactly generated. But $X \times X'$ has neither property. Note that X is an example of a Schwartz space whose dual space X' is not a Schwartz space. □

4. DIFFERENTIABLE FUNCTIONS ON bw* SPACES

In chapter 2 we developed the notion of a bw* space, and emphasized the idea that a bw* space X has two natural locally convex topologies: the original bw* topology, and the Banachable topology E which X has when regarded as the dual space of X'. Since both X and E are D-spaces, it is natural to attempt to relate the notion of differentiability of a map $f : X_1 \to X_2$ to the usual notion of the differentiability of f considered as a map between the associated Banach spaces E_1 and E_2. The former notion will be referred to as "weak differentiability" or simply as "differentiability", and the latter notion as "strong differentiability" (after the weak and strong topologies, respectively).

Weak differentiability will turn out to be a much more restrictive notion than so-called strong differentiability. For instance, if $f : X \to R$ is a weakly C^k map then, since the canonical map $E \to X$ is continuous and linear, $f : E \to R$ is C^k.

On the other hand, let E be an infinite-dimensional Hilbert space, and let X be E with the bounded weak topology. Let $g : E \to R$ be a smooth function such that:

(1) $g(0) \neq 0$.

(2) the support of g is contained in the unit ball of E.

Then $g^{-1}(R - \{0\})$ is non-empty and has compact closure in X, so that $g^{-1}(R - \{0\})$ is not weakly open. Thus g is not even weakly continuous.

We continue to use X to represent a bw* space, E a Banach space (usually the second dual of X), and Z to indicate an arbitrary locally convex space.

<u>4.1 Lemma</u>. Let X be a bw* space. Then, for each $r \in N$, $L^r(X,R)$ is a closed linear subspace of $L^r(E,R)$.

Proof. Obviously $f \in L^r(X,\mathbb{R})$ implies $f \in L^r(E,\mathbb{R})$. Let K be a compact barrel in X. Then the topology on both $L^r(X,\mathbb{R})$ and $L^r(E,\mathbb{R})$ is defined by uniform convergence on K^r, so that $L^r(X,\mathbb{R})$ is topologically embedded in $L^r(E,\mathbb{R})$. Finally, $L^r(X,\mathbb{R})$ is complete by Proposition 3.23, so that $L^r(X,\mathbb{R})$ is a closed subspace of $L^r(E,\mathbb{R})$. □

4.2 Remark. In general, $L^r(X,\mathbb{R})$ is not equal to $L^r(E,\mathbb{R})$. For instance, if $r > 1$, $L^r(X,\mathbb{R}) = L(X,L^{r-1}(X,\mathbb{R}))$. Let K be a compact barrel in X, and assume $f \in L(X,L^{r-1}(X,\mathbb{R}))$. Then $f(K)$ is a compact subset of the Banach space $L^{r-1}(X,\mathbb{R})$. Thus, if $g \in L(E,L^{r-1}(E,\mathbb{R}))$ and g is not completely continuous, then $g \in L^r(E,\mathbb{R})$ but $g \notin L^r(X,\mathbb{R})$.

In the case $r = 1$, note that $L(X,\mathbb{R}) = X'$, and $L(E,\mathbb{R}) = E' = X'''$, so that $L(X,\mathbb{R}) = L(E,\mathbb{R}) \iff E$ is reflexive.

4.3 Notation. If X is a bw* space, U open in X, and $g : U \to L^r(E,\mathbb{R})$ a set mapping, then we will say that g is weakly continuous (resp. C^k) if $g(U) \subset L^r(X,\mathbb{R})$ and $g : U \to L^r(X,\mathbb{R})$ is continuous (resp. C^k).

4.4 Theorem. Let X be a bw* space, U an open subset of X, $f : U \to \mathbb{R}$ a set mapping, and $k \in \mathbb{N}$. Then f is weakly $C^k \iff f$ is strongly C^k and $D^k f$ is weakly continuous for all $1 \le i \le k$.

Proof. Assume f is weakly C^k. Then $D^i f$ is weakly continuous for all $1 \le i \le k$. As we remarked above, since the natural map $E \to X$ is continuous and linear, it is C^∞, so that the closure of C^k maps under composition implies that f is strongly C^k. Conversely, assume that f is strongly C^k and that $D^i f$ is weakly continuous for all $1 \le i \le k$. Since all the conditions on f in this theorem are local in nature, we may assume that U is convex. Let $0 \le r \le k-1$. Since $D^r f$ is a strongly C^1 map from U to $L^r(E,\mathbb{R})$,

$$D^r f(x+y) - D^r f(x) = \int_0^1 D^{r+1} f(x+ty)(y)\,dt.$$

Since $D^{r+1}f$ is weakly continuous, Lemma 3.28 implies that $D^r f$ is weakly C^1, and $D(D^r f) = D^{r+1}f$. Thus f is weakly C^k. □

Before we prove the analogue of the above theorem for maps between bw* spaces, we will make a few remarks about multi-linear maps between bw* spaces. Let $f \in L^r(X_1,X_2)$. Then f maps bounded subsets of X_1^r to bounded subsets of X_2, so $f \in L^r(E_1,E_2)$. The topology on $L^r(E_1,E_2)$ is defined by uniform convergence in E_2 on bounded subsets of E_1^r, and the topology on $L^r(X_1,X_2)$ is defined by uniform convergence in X_2 on bounded subsets of X_1^r. Since the bounded subsets in X_1^r are the same as the bounded subsets in E_1^r, the topology which $L^r(X_1,X_2)$ inherits as a subspace of $L^r(E_1,E_2)$ is at least as strong as the original topology on $L^r(X_1,X_2)$, and is strictly stronger if X_2 is infinite-dimensional. Thus, while we may regard $L^r(X_1,X_2)$ as a subset of $L^r(E_1,E_2)$, we cannot in general regard $L^r(X_1,X_2)$ as a subspace of $L^r(E_1,E_2)$.

<u>4.5 Lemma</u>. $L^r(X_1,X_2)$ is a closed subset of $L^r(E_1,E_2)$.

<u>Proof</u>. Let $\{f_i\}_{i=N}$ be a Cauchy sequence in $L^r(E_1,E_2)$ such that $f_i \in L^r(X_1,X_2)$ for each $i \in N$. Since $L^r(E_1,E_2)$ is complete, there exists $f \in L^r(E_1,E_2)$ such that $f = \lim_{i\to\infty} f_i$, where the limit is taken in $L^r(E_1,E_2)$. Since the topology which $L^r(X_1,X_2)$ inherits as a subset of $L^r(E_1,E_2)$ is stronger than the topology on $L^r(X_1,X_2)$, $\{f_i\}_{i\in N}$ is also a Cauchy sequence in $L^r(X_1,X_2)$. Proposition 3.23 implies that $L^r(X_1,X_2)$ is complete, so there exists $g \in L^r(X_1,X_2)$ such that $g = \lim_{i\to\infty} f_i$, where the limit is taken in $L^r(X_1,X_2)$. To complete the proof, it suffices to show that $f = g$. To see this, let $\ell \in L(X_2,R)$. Then ℓ induces a continuous linear map $L^r(X_1,X_2) \to L^r(X_1,R)$,

$$h \to \ell \circ h$$

so $\ell \circ g = \lim_{i\to\infty} \ell \circ f_i$. Similarly, ℓ induces a continuous linear map $L^r(E_1,E_2) \to L^r(E_1,R)$. and $\ell \circ f = \lim_{i\to\infty} \ell \circ f_i$. But $L^r(X_1,R)$ is a subspace of $L^r(E_1,R)$, so $\ell \circ f = \ell \circ g$. Since this is true for each $\ell \in L(X_2,R)$, we conclude that $f = g$. □

4.6 Definition. Let X_1, X_2 be bw* spaces, U an open subset of X_1, and $f : U \to L^r(E_1, E_2)$ a set mapping. We will say that f is weakly continuous if $f(U) \subset L^r(X_1, X_2)$ and $f : U \to L^r(X_1, X_2)$ is continuous.

Let Z_1 and Z_2 be locally convex spaces. The topology on $L^r(Z_1, Z_2)$ has the property that a subset A of $L^r(Z_1, Z_2)$ is bounded
$\Longleftrightarrow \bigcup_{f \in A} f(B^r)$ is a bounded subset of Z_2 for each bounded subset B of Z_1. In particular, we have the following:

4.7 Lemma. A subset A of $L^r(X_1, X_2)$ is bounded $\Longleftrightarrow$ it is a bounded subset of $L^r(E_1, E_2)$.

4.8 Lemma. Let K_i be a compact barrel in X_i, and let E_i have the associated dual norm, $i = 1,2$. Then, for each $a \geq 0$, $\{f \in L^r(X_1, X_2) \mid \|f\| \leq a\}$ is closed in $L^r(X_1, X_2)$.

Proof. This follows since $ev_w : L^r(X_1, X_2) \to X_2$, $f \to f(w)$ is continuous for each $w \in X_1^r$. □

4.9 Remark. Let X_i, K_i, and E_i be as in the above lemma, and let $\alpha : I \to L^r(X_1, X_2)$ be continuous. Assume that $\|\alpha(t)\| \leq a$ for each $t \in I$. Then Lemma 4.8 implies that $\|\int_0^1 \alpha(t)\,dt\| \leq a$.

Recall the following definition:

4.10 Definition. Let E_1, E_2 be Banach spaces, U an open subset of E_1, $f : U \to E_2$. Then f is said to be C^{1-} if f is locally uniformly Lipschitz, i.e. for each $x \in U$, there exists a neighborhood U_x of x and $a_x > 0$ such that
$$\frac{|f(y_1) - f(y_2)|}{|y_1 - y_2|} \leq a_x$$
for all $y_1 \neq y_2 \in U_x$. We say that f is C^{k-}, $k > 1$, if f is C^{k-1} and $D^{k-1}f$ is C^{1-}.

4.11 Theorem. Let X_1, X_2 be bw* spaces, U an open subset of X_1, and $k \in N$. Assume $f : U \to X_2$ is strongly C^k, and that $D^i f$ is weakly continuous for all $1 \le i \le k$. Then f is weakly C^k.

Proof. It suffices to assume that U is convex. Let $1 \le i \le k$. Then $D^{i-1}f$ is strongly C^1, so

$$D^{i-1}f(x+y) - D^{i-1}f(x) = \int_0^1 D^i f(x+ty)(y)\,dt \quad \text{for all } x, x+y \in U.$$

Lemma 4.5 implies that the integral $\int_0^1 D^i f(x+ty)(y)\,dt$ is the same in $L^{i-1}(X_1,X_2)$ as it is in $L^{i-1}(E_1,E_2)$, so Lemma 3.28 implies that $D^{i-1}f$ is weakly C^1, and that the weak derivative of $D^{i-1}f$ is $D^i f$. Thus f is weakly C^k. □

4.12 Theorem. Let X_1, X_2 be bw* spaces, U an open subset of X_1, and $k \in N$. Assume $f : U \to X_2$ is weakly C^k. Then f is strongly C^{k-}.

Proof. Choose a compact barrel K_i in X_i, $i = 1,2$, and let E_i have the associated norm. Let $x \in U$. There exists $\varepsilon > 0$ such that $B_x(\varepsilon)$, the closed ball around x of radius ε, is contained in U. Assume $1 \le i \le k$. Then $(D^i f)(B_x(\varepsilon))$ is a compact subset of $L^i(X_1,X_2)$, and hence is bounded in $L^i(X_1,X_2)$. Thus there exists $m_i \in N$ such that $\|D^i f(y)\| \le m_i$ for all $y \in B_x(\varepsilon)$. Let $y_1, y_2 \in B_x(\varepsilon)$, and note that

$$\|D^{i-1}f(y_2) - D^{i-1}f(y_1)\| = \left\|\int_0^1 D^i f(ty_2+(1-t)y_1)(y_2-y_1)\,dt\right\|$$

$$\le \left\|\int_0^1 D^i f(ty_2+(1-t)y_1)\,dt\right\| \cdot \|y_2-y_1\| \le m_i\|y_2-y_1\|,$$

so that $D^{i-1}f$ is locally uniformly Lipschitz for all $1 \le i \le k$. Now assume that $1 \le i < k$. Then $D^i f$ is strongly continuous, so $\int_0^1 D^i f(ty_2+(1-t)y_1)(y_2-y_1)\,dt$ exists in $L^{i-1}(E_1,E_2)$, and has the same value as when the integral is taken in $L^{i-1}(X_1,X_2)$. Thus $D^{i-1}f$ is strongly C^1 by 3.28, and the strong derivative of $D^{i-1}f$ is $D^i f$. Thus f is strongly C^{k-1}. Since $D^{k-1}f$ is locally uniformly Lipschitz, f is strongly C^{k-}. □

4.13 Corollary. If X_1 is finite-dimensional, then f is strongly C^∞ <==> f is weakly C^∞.

Proof. Assume f is strongly C^∞. Since X_1 is finite-dimensional, $E_1 = X_1$. Thus f is a C^∞ map from X_1 to E_2. Since the canonical map $E_2 \to X_2$ is C^∞, f is a C^∞ map from X_1 to X_2, i.e. f is weakly C^∞. □

Let f be weakly C^k. Note that $D^i f$ is bounded on each weakly compact subset of U for all $0 \le i \le k$. Thus, loosely speaking, what we have found is a class of differentiable maps whose derivatives are bounded on bounded sets.

The following proposition will be relevant to our discussion of the tangent bundle of a weak manifold:

4.14 Proposition. Let V be an exponential Schwartz space, Z a projective locally convex space, U an open subset of V, and $f : U \to L^r(V,Z)$ a C^k map. Then $ev\circ(f\times I) : U \times V^r \to Z$ is a C^k map.

Proof. Since V is an exponential Schwartz space, V^{r+1} is an exponential Schwartz space, and hence a D-space. Thus it suffices to show that $p_\lambda \circ ev \circ (f\times I)$ is C^k for each semi-norm λ on Z, so we may assume that Z is a normed space, and that $k \in N$. Let $x \in U$. Then there exists a neighborhood U_x of x and a semi-norm ν on V such that $f(U_x) \subset L^r(V_\nu,Z)$ and such that the map $f : U_x \to L^r(V_\nu,Z)$ is C^k. Thus, on $U_x \times V^r$, we have the following factorization of $ev\circ(f\times I)$:

$$U_x \times V^r \xrightarrow{(f,p_\nu^r)} L^r(V_\nu,Z) \times V_\nu^r \xrightarrow{ev} Z ,$$

so that $ev\circ(f\times I)$ is C^k on $U_x \times V^r$ for each $x \in U$. □

We now turn our attention to the question of the existence of smooth partitions of unity on bw* spaces.

Let X be a bw* space, and let U be an open subset of X. Assume that there is a continuous function $f:X \to \mathbb{R}$ such that $U = \{x \in X: f(x) \neq 0\}$. Then, since $f^{-1}(\mathbb{R}-\{0\}) = \bigcup_{n \in N} f^{-1}(\mathbb{R} - (-\frac{1}{n}, \frac{1}{n})) \cap K(n)$, it follows that U must be σ-compact.

What we will next show is a strong converse to this result: namely, if U is σ-compact, then there is a weakly smooth function $f_U:X \to \mathbb{R}$ such that $U = \{x \in X: f_U(x) \neq 0\}$.

Since every open subset of a Bw* space is σ-compact, the existence of smooth partitions of unity on Bw* spaces will follow immediately from this result.

We will break the proof into a series of lemmas. It will be necessary (for the last time in this paper) to make use of the actual weak topology on a bw* space X, as well as the usual, or bounded weak topology.

<u>4.15 Lemma</u>. Let X be a bw* space, let U be a subset of X which is open in the weak topology on X, and let $x \in U$. Then there exists a non-negative C^∞ function $f : X \to \mathbb{R}$ such that $f(x) > 0$, and such that $f^{-1}((0,\infty)) \subset U$.

<u>Proof</u>. There exists $r \in N$, linear functionals $\ell_1,\ldots,\ell_r \in X'$, and real numbers a_i,b_i, $1 \le i \le r$, such that $x \in \bigcap_{i=1}^{r} \ell_i^{-1}((a_i,b_i)) \subset U$. Let $g_i : \mathbb{R} \to \mathbb{R}$ be a smooth non-negative function such that $g_i^{-1}((0,\infty)) = (a_i,b_i)$, and define

$$f = \prod_{i=1}^{r} (g_i \circ \ell_i). \quad \square$$

<u>4.16 Lemma</u>. Let X be a bw* space, let U be a subset of X which is open in the weak topology on X, and let C be a compact subset of U. Then there exists a non-negative C^∞ function $f : X \to \mathbb{R}$ such that $f(x) = 1$ for all $x \in C$, and such that $f^{-1}((0,\infty)) \subset U$.

Proof. For each $x \in C$, choose a smooth non-negative function $f_x : X \to R$ such that $f(x) > 0$, and such that $U_x = f_x^{-1}((0,\infty)) \subset U$. Since $\{U_x | x \in X\}$ is an open cover of C, there exists a finite subcover $U_{x_1},\ldots,U_{x_n}$. Define $h = f_{x_1} + \ldots + f_{x_n}$. Then $h(x) > 0$ for all $x \in C$, and $h^{-1}((0,\infty)) \subset U$. Since C is compact, there exists $\varepsilon > 0$ such that $h(x) > \varepsilon$ for all $x \in C$. Let $g : R \to R$ be a smooth non-decreasing function such that $g(0) = 0$, and such that $g(r) = 1$ for all $r \geq \varepsilon$. Define $f = g \circ h$. □

4.17 Remark. Since the weak topology on X is regular, we may assume that the function f of Lemma 4.16 has been constructed so that the support of f is contained in U.

4.18 Lemma. Let X be a bw* space, let U be an open subset of X, and let C be a compact subset of U. Then there exists a smooth non-negative function $f : X \to R$ such that $f(x) > 0$ for all $x \in C$, and such that $f^{-1}((0,\infty)) \subset U$.

Proof. Let K be a compact barrel in X. By replacing K with a multiple of K if necessary, we may assume $C \subset K$. Since the topology on X is the bounded weak topology, there exists a sequence $U_1, U_2, \ldots$ of weakly open sets such that $U_i \cap K(i) = U \cap K(i)$. Choose a smooth non-negative function $f_1 : X \to R$ such that $f_1(x) > 0$ for all $x \in C$, and such that the support of f_1 is contained in U_1. Let $C_1 = (\text{supp } f_1) \cap K$, and choose a smooth non-negative function $\alpha_1 : X \to R$ such that $\alpha_1(C_1) = 1$, and such that support of α_1 is contained in U_2. Define $f_2 = \alpha_1 f_1$. Then f_2 is a smooth non-negative function on X, $f_2|_K = f_1|_K$, and the support of f_2 is contained in U_2. Proceeding inductively, we may show the existence of a sequence $\{f_n\}_{n \in N}$ of smooth non-negative functions on X such that $f_n|_{K(n-1)} = f_{n-1}|_{K(n-1)}$, and such that the support of f_n is contained in U_n. Define $f : X \to R$ by $f(x) = \lim_{n \to \infty} f_n(x)$ for each $x \in R$. f is obviously non-negative, $f(x) > 0$ for all $x \in C$, and $f^{-1}((0,\infty)) \subset U$. Since $X = \varinjlim_{n \in N} K(n)$, f is continuous. So we need only show that f is smooth. For each $r \in N$, define $g_r : X \to L^r(X,R)$ by $g_r(x) = \lim_{n\to\infty} D^r f_n(x)$. Since $f_{n+m}|_{K(n)} = f_n|_{K(n)}$ for all $m \in N$, g_r obviously exists and is continuous for each $r \in N$. Equally obviously,

$$f(x+y) - f(x) = \int_0^1 g_1(x+ty)(y)\,dt \quad \text{for all } x,y \in X,$$

so f is C^1 and $Df = g$. Similarly, for each $r \in N$,

$$g_r(x+y) - g_r(x) = \int_0^1 g_{r+1}(x+ty)(y)\,dt,$$

so that g_r is C^1 and $Dg_r = g_{r+1}$. Thus f is C^∞. □

4.19 Theorem. Let X be a bw* space, and let U be an open subset of X such that U is σ-compact. Then there exists a smooth non-negative function $f : X \to R$ such that $U = f^{-1}((0,\infty))$.

Proof. Let $U = \bigcup_{i\in N} C_i$, C_i compact. For each $i \in N$, let $f_i : X \to R$ be a smooth non-negative function such that $f_i(x) > 0$ for all $x \in C_i$, and such that $f_i^{-1}((0,\infty)) \subset U$. Choose a compact barrel K in X, and use K to determine norms on the spaces $L^r(X,R)$ of multi-linear maps. For each $i \in N$, choose an integer $n_i > 0$ such that $n_i > \sup\{|f_i(x)|, \|Df_i(x)\|, \ldots, \|D^i f_i(x)\| \mid x \in K(i)\}$, and define $f = \sum_{i=1}^{\infty} \frac{f_i}{2^i n_i}$. Then f has the required properties. □

4.20 Corollary. Let X be a Bw* space, U an open subset of X. Then there exists a smooth non-negative function $f : X \to R$ such that $U = f^{-1}((0,\infty))$.

Lemma 4.18 is the last place that we shall see the weak topology as well as the bounded weak topology on a bw* space. Thus we return to our previous convention of referring to the usual (bounded weak) topology on a bw* space X as the "weak" topology, and to the norm topology on the second dual of X as the "strong" topology.

This paper will be concerned almost exclusively with bw* spaces from now on, so we will not prove results about other classes of D-spaces. However, we will mention that 4.20 admits a generalization to duals of Fréchet Schwartz spaces. Specifically, let V be a Fréchet Schwartz space, V' its dual space. If U is an open subset of V', then U is σ-compact. Furthermore, there exists a smooth function $f_U: V' \to \mathbb{R}$ such that $U = f_U^{-1}((0,\infty))$.

5. AN INVERSE FUNCTION THEOREM

5.1 Definition. Let Z be a locally convex space. We define $\mathrm{Aut}(Z)$ to be the subspace of $L(Z,Z)$ whose underlying set is $\{f \in L(Z,Z) \mid f$ is bijective and $f^{-1} \in L(Z,Z)\}$.

A natural question to consider at this point is whether the inverse function theorem for C^1 maps between Banach spaces can be generalized to C^1 maps between D-spaces. It should not be too surprising that this is a much more difficult problem than the corresponding problem for Banach spaces.

The first thing to note is that any incomplete normed space is a D-space. Since the inverse function theorem does not hold for incomplete normed spaces, the answer to our question is in general: No.

Let V be a D-space, U a neighborhood of zero in V, and $f : U \to V$ a C^1 map such that $Df(0) \in \mathrm{Aut}(V)$. If the inverse function theorem holds for C^1 endomorphisms of V, then $W = \{x \in V \mid Df(x) \in \mathrm{Aut}(V)\}$ is a neighborhood of zero, and the map $Df^{-1} : W \to \mathrm{Aut}(V)$ is continuous. Thus, to show that the inverse function theorem does not hold for C^1 endomorphisms of V, it suffices to produce a C^1 map $f : V \to V$ such that $Df(0) \in \mathrm{Aut}(V)$ and such that $\{x \in V \mid Df(x) \in \mathrm{Aut}(V)\}$ is not a neighborhood of zero.

Of course, such a map cannot be constructed if V is a Banach space. For, if V is a Banach space, then recall (see [26] or [30]) that $\mathrm{Aut}(V)$ is an open subspace of $L(V,V)$ and that the map

$$\begin{aligned} \mathrm{inv}:\mathrm{Aut}(V) &\longrightarrow \mathrm{Aut}(V) \\ \ell &\longrightarrow \ell^{-1} \end{aligned}$$

is smooth (even analytic). Thus, if $g:V \to V$ is C^1, then $Dg:V \to L(V,V)$ is continuous, so that $\{x \in V \mid Dg(x) \in \mathrm{Aut}(V)\}$ is open, and $(Dg)^{-1}$ is a continuous map.

However, the above argument holds only for Banach spaces and certain other normed spaces, as observed by J. N. Vladimirskii in [46]. Specifically, he and R. Bonic (unpublished) independently established the following result:

5.2 Lemma. Let Z be a locally convex space which is not normable. Then $Aut(Z)$ is not open in $L(Z,Z)$.

Proof. We will show that the identity automorphism is not an interior point of $Aut(Z)$.

For each semi-norm λ on Z, and nonempty bounded closed convex balanced subset B of Z, consider the semi-norm $\mu = \mu(B,\lambda)$ on $L(Z,Z)$ defined by $\mu(g) = \sup\{\lambda(g(z)) \mid z \in B\}$ for each $g \in L(Z,Z)$. These semi-norms generate the topology on $L(Z,Z)$. Now, for each semi-norm μ, let $U_\mu = \{g \in L(Z,Z) \mid \mu(Id-g) < 1\}$. The collection of open sets $\{U_\mu\}$ is a base for the neighborhood system at the identity automorphism Id. Thus, to show that $Aut(Z)$ is not open, it suffices to show that, for each semi-norm μ, there is an element $g_\mu \in U_\mu$ such that $g_\mu \notin Aut(Z)$.

So let $\mu = \mu(B,\lambda)$ be one of the above semi-norms. Since Z is not normable, $\lambda^{-1}([0,1))$ is not bounded in Z. Thus there exists $x \in \lambda^{-1}([0,1))$ such that $x \notin B$. Since B is closed, convex and balanced, there exists $\ell \in Z'$ such that $\ell(x) = 1$ and such that $|\ell(z)| \leq 1$ for each $z \in B$ (see [38], p. 108). Define $h \in L(Z,Z)$ by $h(z) = (\ell(z))\cdot x$ for each $z \in Z$, and define g_μ by $g_\mu = Id-h$. Since $\mu(h) \leq \lambda(x) < 1$, we see that $g_\mu \in U_\mu$. But $g_\mu(x) = x - h(x) = 0$, which implies that the kernel of g_μ is non-zero, and in particular that $g_\mu \notin Aut(Z)$. □

Thus if V is a D-space which is not normable, it is fairly reasonable to suspect the existence of a map f of the type discussed above.

This does not mean, though, that we cannot prove an inverse function theorem for C^1 endomorphisms of V; it simply means that we have to look for special subclasses of C^1 endomorphisms of V, i.e., maps which have some extra condition to compensate for the fact that $\mathrm{Aut}(V)$ is not an open subspace of $L(V,V)$.

This is what we will do for bw* spaces. We will construct a smooth map f from the weak separable Hilbert space X to itself for which $\{x \in X \mid Df(x) \in \mathrm{Aut}(X)\}$ is not open in X, and then we will find a very useful class of maps between bw* spaces for which we CAN prove an inverse function theorem.

First, though, we will show that the C^k inverse function theorem follows from the C^1 case - i.e. we will show that a C^k map which is a C^1 isomorphism, is also a C^k isomorphism.

5.3 Lemma. Let V be a D-space, and $U \subseteq V$ a neighborhood of 0. Let $g : U \to L(V,V)$ and $h : U \to L(V,V)$ be maps such that $g(0) = h(0) = 0$. Assume that g is continuous, and that h is differentiable at 0. Then $g \cdot h$ is $o(\|y\|)$.

Proof. Let $\mu(B,\lambda)$ be one of the basic semi-norms on $L(V,V)$, where λ is a semi-norm on V, B a bounded subset of V, and $\mu(B,\lambda)(f) = \sup\{\lambda(f(x)) \mid x \in B\}$ for each $f \in L(V,V)$. The map $g_\lambda = p_\lambda \circ g : U \to L(V,V_\lambda)$ is continuous, so there exists a semi-norm ν on V, and a neighborhood U_0 of 0, such that $g_\lambda(U_0) \subset L(V_\nu,V_\lambda)$ and such that the map $g_\lambda : U_0 \to L(V_\nu,V_\lambda)$ is continuous. Now, let us consider the semi-norm $\mu(B,\nu)$ on $L(V,V)$: since h is differentiable at 0, there exists a semi-norm α on V and a neighborhood $W \subseteq U_0$ such that $\mu(B,\nu)(h(y)) \leq \alpha(y)$ for all $y \in W$. Given $\varepsilon > 0$, choose $U_\varepsilon \subseteq W$ such that $\|g_\lambda(x)\| \leq \varepsilon$ for all $x \in U_\varepsilon$, where this last norm is the norm on $L(V_\nu,V_\lambda)$. Then $\mu(B,\lambda)(g \cdot h(y)) \leq \varepsilon\alpha(y)$, which means that $g \cdot h$ is tangent to 0 at 0. □

5.4 Lemma. Let V be a projective D-space, U an open subset of V, $k \geq 1$, and $g : U \to \mathrm{Aut}(V)$ a C^k map. Assume in addition that g^{-1} is continuous. Then g^{-1} is C^k.

Proof. Let $x \in U$. We will first show that g^{-1} is differentiable at x:

$$g^{-1}(x+v) - g^{-1}(x) = g^{-1}(x+v)(g(x) - g(x+v))g^{-1}(x)$$

$$= (g^{-1}(x+v) - g^{-1}(x))(g(x) - g(x+v))g^{-1}(x)$$
$$+ g^{-1}(x)(g(x) - g(x+v))g^{-1}(x) .$$

Now the map $v \to g(x) - g(x+v)$ is differentiable at 0, so the product $(g^{-1}(x+v) - g^{-1}(x))(g(x) - g(x+v))$ is $o(\|v\|)$ by the previous lemma. Since the map $L(V,V) \to L(V,V)$ is linear,

$$\ell \to \ell \cdot g^{-1}(x)$$

the first term in the above sum is $o(\|v\|)$. Since g is differentiable at x, $g(x+v) - g(x) = Dg(x)(v) + r_x(v)$, where $Dg(x) \in L(V,L(V,V))$ and r_x is $o(\|v\|)$. Thus

$$g^{-1}(x)(g(x) - g(x+v))g^{-1}(x) = -g^{-1}(x)\cdot Dg(x)(v)\cdot g^{-1}(x)$$
$$-g^{-1}(x)\cdot r_x(v)\cdot g^{-1}(x)$$

Since the map $L(V,V) \to L(V,V)$ is linear,
$$\ell \to g^{-1}(x)\cdot \ell \cdot g^{-1}(x)$$
$g^{-1}(x)\cdot r_x(v)\cdot g^{-1}(x)$ is $o(\|v\|)$, which implies that g^{-1} is differentiable at x, and $D(g^{-1})(x) = g^{-1}(x)\cdot Dg(x)\cdot g^{-1}(x)$. Since g^{-1} and Dg are both continuous, this implies that $D(g^{-1})$ is continuous, which in turn implies that g^{-1} is C^1. By induction, we conclude that g^{-1} is C^k. □

Note again that, if V is a Banach space, then g^{-1} is automatically continuous, and the above lemma follows from the fact that inv : $\mathrm{Aut}(V) \to \mathrm{Aut}(V)$ is C^∞. Now, if V is a more general projective D-space, g^{-1} might **not** be continuous. However, we have shown that, IF g^{-1} IS continuous, then it still follows that g^{-1} is C^k. The next theorem is actually a corollary of this result.

<u>5.5 Theorem.</u> Let V be a projective D-space, U and W open subsets of V, $k > 1$, and $f : U \to W$ a C^k bijection. Assume in addition that f^{-1} is C^1. Then f^{-1} is C^k.

Proof. Obviously, for each $x \in U$, $Df(x) \in \mathrm{Aut}(V)$. In fact, $(Df(x))^{-1} = (D(f^{-1}))(f(x))$, so that $(Df)^{-1} = (D(f^{-1})) \circ f$. This implies that $(Df)^{-1}$ is continuous, so that $(Df)^{-1}$ is C^{k-1} by the previous lemma. But we may write $D(f^{-1}) = (Df)^{-1} \circ f^{-1}$. Since $(Df)^{-1}$ is C^{k-1} and f^{-1} is C^1, $D(f^{-1})$ is C^1, which implies that f^{-1} is C^2. By induction, we conclude that $D(f^{-1})$ is C^{k-1}, and hence that f^{-1} is C^k. □

From now on, we will restrict our attention to bw* spaces. We first present an example of a weakly smooth map f from the weak separable Hilbert space X to itself for which $Df(0)$ is the identity map, but for which $\{x \in X \mid Df(x) \in \mathrm{Aut}(X)\}$ is not a neighborhood of 0.

5.6 Example. Let H be a separable infinite-dimensional Hilbert space, with orthonormal basis $e_1, e_2, \ldots$ Let $\rho : \mathbb{R} \to [0,1]$ be a smooth function such that $\rho(r) = 1$ for all $r \le 1/2$ and $\rho(r) = 0$ for all $r \geq 1$. For each $x = (x^1, x^2, \ldots) \in H$, define $f : H \to H$ by $f(x) = \sum_{i=1}^{\infty} x^i \rho(x^i) e_i$. Then f is weakly smooth and $Df(0) = \mathrm{Id}$. But $Df(e_i) \notin \mathrm{Aut}(H)$ for any $i \in N$, so that $\{x \in H \mid Df(x) \in \mathrm{Aut}(X)\}$ is not a neighborhood of 0.

Proof. It suffices to show that f is weakly smooth. For then it is immediate that $Df(e_i)(e_i) = 0$, so that $Df(e_i) \notin \mathrm{Aut}(X)$ for any $i \in N$. And, since f is the identity on $B_0(1/2)$, $Df(0)$ is obviously the identity.

We will proceed by showing that f is strongly smooth, and that $D^i f$ is weakly continuous for all $i \in N$. To see that f is strongly C^∞, note that $f = \mathrm{Id}$ on $B_0(1/2)$, so f is smooth on $B_0(1/2)$. If $\|x\| \geq 1/2$, there exists $m \in N$ such that $|x^i| < 1/4$ for all $i \geq m$. Thus, if $y \in B_x(1/4)$, $(f(y))^i = y^i$ for all $i \geq m$. This implies that, on $B_x(1/4)$, f can be represented as the identity plus a finite number of strongly smooth maps, which implies that f is strongly C^∞ on $B_x(1/4)$. Thus f is strongly C^∞. To see that each derivative of f is weakly continuous, we first compute the

derivatives: it is easy to see that
$Df_x(v) = \Sigma_{i=1}^{\infty} e_i(v^i \rho(x^i) + x^i v^i \rho'(x^i))$, and in general that

$$D^r f_x(v_1,\dots,v_r) = \Sigma_{i=1}^{\infty} e_i(r v_1^i \dots v_r^i \rho^{r-1}(x^i) + x^i v_1^i \dots v_r^i \rho^r(x^i)).$$

Since the derivatives of ρ are bounded functions, the derivatives of f are bounded on bounded subsets of X. Let $r \in N$. Since the topology on $L^r(X,X)$ is the compact-open topology (i.e. $L^r(X,X) \subset C^0(X^r,X)$), to check the weak continuity of $D^r f$, it suffices to show that the map

$$ev\circ(D^r f \times id) : X^{r+1} \longrightarrow X \qquad \text{is weakly}$$
$$(x,x_1,\dots,x_r) \longrightarrow D^r f_x(x_1,\dots,x_r)$$

continuous. Since X^{r+1} is compactly generated, it suffices to check the continuity on compact subsets of X^{r+1}. Let C be a compact subset of X^{r+1}. Since $D^r f$ is bounded on bounded sets, $A = ev\circ(D^r f \times id)(C)$ is bounded on X, i.e. $\overline{A}$ is compact. Since the linear functionals $\{e_i\}_{i\in N}$ separate the points of A, these functionals generate the topology on A. Thus, to check that $ev\circ(D^r f \times id)$ is continuous on C, it suffices to check that $e_i\circ ev\circ(D^r f \times id)$ is continuous on C for each $i \in N$. Now $e_i\circ ev\circ(D^r f \times id) = ev\circ(D^r f_i \times id)$, where f_i is the ith coordinate of f. But f_i is obviously weakly C^∞, so that $D^r f_i$ is weakly continuous, and hence $ev\circ(D^r f_i \times id)$ is weakly continuous. Thus f is weakly C^∞. □

If we hope to establish an inverse function theorem for bw* spaces, we must restrict our attention to classes of maps which satisfy some kind of uniformity condition on the first derivative. Now, although there are many inequivalent continuous semi-norms on a bw* space, none of these has any distinguishing properties. However, there is one class of distinguished (though not continuous) norms on a bw* space X - namely the norms which define the associated Banach space structure on X. Thus it is natural to consider weakly C^1 maps whose first derivatives are "regulated" in some sense by the strong norm on X. The specific property which we will be interested in suggests itself if one examines the global-analytic treatment of the calculus of variations due to Palais and Uhlenbeck. We will begin with a generalization of the basic property of D-spaces.

5.7 Lemma. Let $X, X_1, \ldots, X_r$ be bw* spaces, U an open subset of X, E a normed space, and $g : U \to L^r(X_1,\ldots,X_r;E)$ a continuous map. Then, for each compact subset $C \subset U$, there exists a neighborhood W of C in U and semi-norms ν_i on X_i, $1 \le i \le r$, such that $g(W) \subset L^r((X_1)_{\nu_1},\ldots,(X_r)_{\nu_r};E)$ and such that the map $g: W \to L^r((X_1)_{\nu_1},\ldots,(X_r)_{\nu_r};E)$ is continuous.

Proof. Same as 3.46. □

5.8 Definition. Let X_1, X_2, X_3 be bw* spaces, U an open subset of X_1 and $g : U \to L(X_2,X_3)$ a continuous map. We will say that g is $\mathcal{U}$-regulated if, for each compact subset $C \subset U$, there exist continuous semi-norms λ on X_1 and ν on X_2 such that $\|(g(y)-g(x))(z)\| \le \lambda(y-x)\|z\| + \|y-x\|\nu(x)$ for each $x,y \in C$, $z \in X_2$ (where $\|v\|$ denotes the norm of v in E_i for each $v \in X_i$).

Note that, if $X_1 = X_2$, we may replace λ and ν with $(\lambda+\nu)$, and so may assume that $\lambda = \nu$. When we deal with the case $X_1 = X_2$, we will make this simplification without comment.

5.9 Lemma. Let $g : U \to L(X_2,X_3)$ be $\mathcal{U}$-regulated. Then g is strongly continuous, and, in fact, is Lipschitz in the strong norm on each compact subset of U.

Proof. Let C be a compact subset of U. There exist semi-norms λ on X_1 and ν on X_2 such that $\|(g(u)-g(v))(z)\| \le \lambda(u-v)\|z\| + \|u-v\|\nu(z)$ for all $u,v \in C$, $z \in X_2$. Choose $n \in N$ such that $\lambda(y) \le n\|y\|$ for all $y \in X_1$ and $\nu(z) \le n\|z\|$ for all $z \in X_2$. Then $\|g(u)-g(v)\| \le 2n\|u-v\|$, and so we conclude that g is Lipschitz on C. $\square$

5.10 Proposition. Let X, X_1, X_2, X_3 be bw* spaces, U open in X, $g : U \to L(X_1, X_2)$ and $h : U \to L(X_2, X_3)$ $\mathcal{U}$-regulated. Then $h \cdot g : U \to L(X_1, X_3)$ is $\mathcal{U}$-regulated.

Proof. Let C be a compact subset of U, and let $x, y \in C$, $z \in X_1$. $\|(h(y)g(y)-h(x)g(x))(z)\| \leq \|h(y)(g(y)-g(x))(z)\| + \|(h(y)-h(x))g(x)(z)\|$. First, there exist positive constants r, s such that $\|h(y)\| \leq r$ and $\|g(y)\| \leq s$ for all $y \in C$. By definition of $\mathcal{U}$-regulated, there exist semi-norms λ_1 on X and μ on X_2 such that $\|(h(y)-h(x))(w)\| \leq \lambda_1(y-x)\|w\| + \|y-x\|\mu(w)$. Also, there exist semi-norms λ_2 on X and β on X_1 such that $\|(g(y)-g(x))(z)\| \leq \lambda_2(y-z)\|z\| + \|y-x\|\beta(z)$. Finally, there exists a semi-norm α on X_1 such that $g_\mu(C) \subset L_\alpha(X_1, (X_2)_\mu)$ and $g_\mu : C \to L_\alpha(X_1, (X_2)_\mu)$ is continuous. By replacing α with a multiple of α if necessary, we may assume $\|g_\mu(x)\| \leq 1$ for all $x \in C$. Thus

$$\|(h(y)g(y)-h(x)g(x))(z)\| \leq r\lambda_2(y-x)\|z\| + r\|y-x\|\beta(z)$$
$$+\lambda_1(y-x)\cdot s\cdot\|z\| + \|y-x\|\cdot\alpha(z).$$

Letting $\lambda = r\lambda_2 + s\lambda_1$ and $\nu = r\beta + \alpha$, we see that $h \cdot g$ is $\mathcal{U}$-regulated. $\square$

5.11 Proposition. Let X_1, X be bw* spaces, U open in X_1, $g : U \to L(X, X)$ a $\mathcal{U}$-regulated map. Assume that $g(U) \subset \mathrm{Aut}(X)$ and g^{-1} is continuous. Then g^{-1} is $\mathcal{U}$-regulated.

Proof. Let C be a compact subset of U, $x, y \in C$, $z \in X$. Then $\|(g^{-1}(y)-g^{-1}(x))(z)\| = \|g^{-1}(y)(g(x)-g(y))g^{-1}(x)(z)\|$. First, since C is weakly compact, there exists $n > 0$ such that $\|g^{-1}(y)\| \leq n$ for all $y \in C$. Since g is $\mathcal{U}$-regulated, there exist semi-norms λ on X_1 and ν on X_2 such that $\|(g(y)-g(x))(w)\| \leq \lambda(y-x)\|w\| + \|y-x\|\mu(w)$. Consider the map $(g^{-1})_\mu : U \to L(X, X_\mu)$. Since C is compact, there exists a semi-norm ν on X such that $(g^{-1})_\mu(C) \subset L_\nu(X, X_\mu)$ and such that

$(g^{-1})_\mu : C \to L_\nu(X,X_\mu)$ is continuous. By replacing ν with a multiple of ν if necessary, we may assume that $\|(g^{-1})_\mu(x)\| \le 1$ for all $x \in C$. Thus $\|(g^{-1}(y)-g^{-1}(x))(z)\| = \|g^{-1}(y)(g(x)-g(y))g^{-1}(x)(z)\|$ $\le n\|(g(x)-g(y))g^{-1}(x)(z)\| \le n\lambda(y-z)\|g^{-1}(x)(z)\| + n\|y-z\|\mu(g^{-1}(x)(z))$ $\le n^2\lambda(y-x)\|z\| + n\|y-x\|\nu(z)$. Letting $\alpha = n^2\lambda$ and $\beta = n\nu$, we have that $\|(g^{-1}(y)-g^{-1}(x))(z)\| \le \alpha(y-x)\|z\| + \|y-x\|\beta(z)$. □

We have now proved several results about the behavior of a continuous map $g : U \to \mathrm{Aut}(X)$ under the assumption that g^{-1} is continuous, but we have not yet determined any conditions on g which will ensure the continuity of g^{-1}. We will now remedy this omission.

First, we will show that the inverse map $\mathrm{inv}:\mathrm{Aut}(X) \to \mathrm{Aut}(X)$ is not continuous for any bw* space X. Then we will show that the inverse map is nevertheless continuous on certain subsets of $\mathrm{Aut}(X)$ which are sufficiently large for our purposes.

5.12 Proposition. Let X be an infinite-dimensional bw* space. Then $\mathrm{inv}:\mathrm{Aut}(X) \to L(X,X)$ is not continuous.

Proof. Recall that, if $E = X''$, a subset of $L(X,X)$ is bounded if and only if it is also bounded in $L(E,E)$. Thus, to show that $\mathrm{inv}:\mathrm{Aut}(X) \to L(X,X)$ is not continuous, it suffices to exhibit a sequence $\{f_n\}_{n\in N}$ in $\mathrm{Aut}(X)$ with the following properties:

(1) $f_n \to \mathrm{Id}$

(2) $\|f_n^{-1}\| \to \infty$.

Note that, since the sequence $\{f_n^{-1}\}_{n\in N}$ is not bounded, it certainly cannot converge to a limit in $L(X,X)$, and thus the inverse map cannot be continuous on $\mathrm{Aut}(X)$.

To construct a sequence in $L(X,X)$ with the above properties, let $\{v_n\}_{n\in N}$ be a sequence of vectors in X such that $\|v_n\| = 1$ for all $n \in N$ and such that $\lim_{n\to\infty} v_n = 0$. The existence of a sequence of vectors in X with these properties was recently established by Nussenweig (for the case of X a Bw* space, in which the compact subsets are metrizable, the existence of this sequence is easily established by elementary methods).

For each $n \in N$, let $\ell_i \in L(X,\mathbb{R})$ be a functional such that $\ell_i(v_i) = 1$ and such that $\|\ell_i\| = 1$. Then, if we define $g_n : X \to X$ by $g_n(v) = (\ell_n(v))\cdot v_n$ for each $v \in X$, $n \in N$, it follows that $\|g_n\| = 1$ and $\lim_{n\to\infty} g_n = 0$.

Finally, define $f_n : X \to X$ by $f_n = Id - (\frac{n-1}{n})g_n$. Note that $\|Id - f_n\| = (n-1)/n < 1$ for each $n \in N$, which implies that $f_n \in Aut(X)$ for each $n \in N$. Furthermore, $\lim_{n\to\infty} f_n = Id - \lim_{n\to\infty} g_n = Id$. And $\|f_n^{-1}\| \geq \|f_n^{-1}(v_n)\| = n$, so $\lim_{n\to\infty}\|f_n^{-1}\| = \infty$. □

5.13 Remark. Let H be a separable infinite-dimensional Hilbert space with orthonormal basis $e_1, e_2, \ldots$. For each $i \in N$, define $\ell_i \in Aut(H)$ by

$$\ell_i(e_j) = \begin{cases} e_j, & \text{if } j \neq i, \\ (1/i)e_i, & \text{if } j = i. \end{cases}$$

Then, using the techniques employed in Example 5.6, it is easy to construct a weakly C^∞ map $f : H \to H$ such that:

(1) $Df(H) \subset Aut(H)$

(2) $Df(0) = Id$

(3) $Df(e_i) = \ell_i$

(4) $f : H \to H$ is a weak homeomorphism.

Note that, for this example, $(Df)^{-1}$ is not weakly continuous, so that f^{-1} cannot be weakly C^1. □

5.14 Theorem. Let Z be a locally convex space, and let A be a subset of $Aut(Z)$ such that $\{\ell^{-1} | \ell \in A\}$ is a bounded subset of $L(Z,Z)$. Then $inv : A \to L(Z,Z)$ is continuous.

Proof. Let $\mu = \mu(\lambda, B)$ be one of the generating semi-norms for the topology on $L(Z,Z)$, where λ is a semi-norm on Z, B is a bounded subset of Z, and $\mu(\ell) = \sup\{\lambda(\ell(v)) \mid v \in B\}$ for each $\ell \in L(Z,Z)$. Also, let $\ell_0 \in A$ and let $\varepsilon > 0$. Then, to prove the theorem, it suffices to demonstrate the existence of a neighborhood U of ℓ_0 in $L(Z,Z)$ such that, for each $\ell \in U \cap A$, $\mu(\ell^{-1}-\ell_0^{-1}) < \varepsilon$.

Define a subset D of Z by $D = \{\ell^{-1}(v) \mid \ell \in A, v \in B\}$. Then, since $\{\ell^{-1} \mid \ell \in A\}$ is bounded in $L(Z,Z)$, it follows that D is bounded in Z. Define a semi-norm ν on Z by $\nu = \lambda \circ \ell_0^{-1}$.

Now, consider the semi-norm $\tilde{\mu} = \tilde{\mu}(\nu, D)$ on $L(Z,Z)$ defined by $\tilde{\mu}(\ell) = \sup\{\nu(\ell(v)) \mid v \in D\}$. Since subtraction is a continuous operation on $L(Z,Z)$, there exists a neighborhood U of ℓ_0 in $L(Z,Z)$ such that $\tilde{\mu}(\ell-\ell_0) < \varepsilon$. Thus we conclude:

$$\mu(\ell^{-1}-\ell_0^{-1}) = \mu(\ell_0^{-1}-\ell^{-1}) = \sup\{\lambda((\ell_0^{-1}-\ell^{-1})(v)) \mid v \in B\}$$

$$= \sup\{\lambda(\ell_0^{-1}(\ell-\ell_0)(\ell^{-1}(v))) \mid v \in B\} \le \sup\{\lambda(\ell_0^{-1}(\ell-\ell_0)(v)) \mid v \in D\}$$

$$= \sup\{\nu((\ell-\ell_0)(v)) \mid v \in D\} = \tilde{\mu}(\ell-\ell_0) < \varepsilon. \quad \square$$

5.15 Lemma. Let X_1 and X be bw* spaces, U open in X_1, and $g: U \to L(X,X)$ a $\mathcal{U}$-regulated map. Assume that C is a compact subset of U, and that $x_0 \in C$ is a point for which $g(x_0) \in \operatorname{Aut}(X)$. Then there exists a neighborhood V of x_0 in C such that $g(V) \subset \operatorname{Aut}(X)$, and furthermore such that $\{(g(x))^{-1} \mid x \in V\}$ is bounded in $L(X,X)$.

Proof. Since g is continuous, it follows that $g(C)$ is compact, and hence bounded in $L(X,X)$. Thus there exists a positive integer $n_0 \in N$ such that $\|x\| \le n_0$ and $\|g(x)\| \le n_0$ for all $x \in C$.

For each $x \in C$, $z \in X$, consider $\|(g(x) - g(x_0))(g(x) - g(x_0))(z)\|$. Since g is $\mathcal{U}$-regulated, there exists semi-norms λ on X_1 and μ on X such that

$$\|(g(x) - g(x_0))(g(x) - g(x_0))(z)\| \le \lambda(x-x_0)\|(g(x) - g(x_0))(z)\|$$

$$+ \|x-x_0\|\mu((g(x) - g(x_0))(z))$$

for each $x \in C$, $z \in X$.

Since $p_\mu \circ g: C \to L(X, X_\mu)$ is continuous, there exists a semi-norm ν on X and a neighborhood W of 0 in U such that $p_\mu \circ g(W) \subset L_\nu(X, X_\mu)$. Furthermore, by multiplying ν by a constant if necessary, we may assume that $\nu(z) \leq \|z\|$ for all $z \in X$.

Let $\tilde{\nu}$ be the induced norm on $L_\nu(X, X_\mu)$, and let $\varepsilon \in (0, 1/2)$.

Choose a neighborhood A of 0 in W such that $\tilde{\nu}(g(x) - g(x_0)) < \varepsilon/4n_0$ for all $x \in A$.
Let $Y = \{x \in X_1 | \lambda(x - x_0) < \varepsilon/4n_0\}$, and let $V = C \cap A \cap Y$. Then, for all $x \in V$,

$$\|(g(x) - g(x_0))(g(x) - g(x_0))(z)\| < (\varepsilon/4n_0)(2n_0)\|z\| +$$

$$+(2n_0)(\varepsilon/4n_0)\nu(z) \leq \varepsilon\|z\| < (\frac{1}{2})\|z\| .$$

Now, the remaining computations will be simpler if we make the assumption that $g(x_0) = Id$. With this assumption, I claim that $\|g(x)(z)\| \geq (1/6n_0)\|z\|$ for all $x \in V$, $z \in X$. For, if there were $x_1 \in V$, $z_1 \in X$ such that $\|g(x_1)(z_1)\| < (1/6n_1)\|z_1\|$, then

$$\|(g(x_1) - Id)\ (g(x_1) - Id)(z_1)\| = \|z_1 - 2g(x_1)(z_1) + (g(x_1))^2(z_1)\| \geq$$

$$\geq \|z_1\| - 2\|g(x_1)(z_1)\| - n_0\|g(x_1)(z_1)\| >$$

$$> \|z_1\| - 2(1/6n_0)\|z_1\| - n_0(1/6n_0)\|z_1\| \geq$$

$$\geq \|z_1\| - (1/3)\|z_1\| - (1/6)\|z_1\| \geq (1/2)\|z_1\| .$$

But this would contradict the previous inequality, and so we have proved that $\|g(x)(z)\| \geq (1/6n_0)\|z\|$ for all $x \in V$, $z \in X$.

Both conclusions in the statement of this lemma will now follow if we can show that $g(x)$ is surjective for each $x \in V$. To see this, assume to the contrary that x_2 is a point in V for which $g(x_2)$ is not surjective. From what we already have shown, it follows that $g(x_2)(X)$ is a closed subspace of X. Thus there exists a vector $z_0 \in X/g(x_2)(X)$ such that $\|z_0\| \in (1/2, 1)$. Let $z_1 \in X$ be a vector which projects to z_0 such that $\|z_1\| = 1$. Then $\|z_1 - z\| > 1/2$ for all $z \in g(x_2)(X)$. But let $y = 2z_1 - g(x_2)(z_1)$.

Then $\|z_1 - g(x_2)(y)\| = \|z_1 - 2g(x_2)(z_1) + (g(x_2))^2(z_1)\| = \|(g(x_2) - Id)^2(z_1)\| < 1/2$. But this is a contradiction, and so we conclude that $g(x)$ must be surjective for all $x \in V$. □

5.16 Theorem. Let X_1 and X be bw* spaces, U an open subset of X_1, and $g:U \to L(X,X)$ a $\mathfrak{U}$-regulated map. Then $W = \{x \in U | g(x) \in Aut(X)\}$ is an open subset of X_1, and $g^{-1}:W \to Aut(X)$ is continuous.

Proof. For each compact subset C of U, Lemma 5.15 implies that $C \cap W$ is an open subset of C. Since U is compactly generated, this implies that W is open in U, and hence also in X_1.

Now let C be a compact subset of W. Then, for each $x \in C$, there is a neighborhood V_x of x in C such that $\{(g(y))^{-1} | y \in V_x\}$ is bounded in $L(X,X)$. Since C is compact, we can cover C with a finite collection of such neighborhoods, which implies that $\{(g(x))^{-1} | x \in C\}$ is bounded. But then Lemma 5.14 implies that $g^{-1}:C \to Aut(X)$ is continuous. Since W is compactly generated, it follows immediately that $g^{-1}:W \to Aut(X)$ must be continuous. □

The preceding theorem is the first step in the establishment of our inverse function theorem. Before proceeding with the remaining steps, we present a simple, but exceedingly useful, criterion for a map to be $\mathfrak{U}$-regulated, and a simple example. More sophisticated criteria will be presented in the next chapter.

5.17 Proposition. Let X_1, X_2, X_3 be bw* spaces, U an open subset of X_1, and $g:U \to L(X_2,X_3)$ a C^1 map. Assume in addition that, for each compact subset $C \subset U$, there exist semi-norms λ on X_1 and ν on X_2 such that $\|Dg_x(y,z)\| \le \lambda(y)\|z\| + \|y\|\nu(z)$ for all $x \in C$, $y \in X_1$, $z \in X_2$. Then g is $\mathfrak{U}$-regulated.

Proof. Since C is compact, there exists a convex balanced neighborhood W of 0 in X_1 such that $C + \overline{W} \subset U$. Let μ be the

Minkowski functional associated to $\overline{W}$. Since C is compact, there exists an integer $n \in N$ such that $C \subset B_0(n)$. Let $D = \overline{W} \cap B_0(2n)$, so that D is compact, and let $A = C + D$. Then A is compact. Furthermore, if $x,y \in C$ and $\mu(x-y) \leq 1$, then $x-y \in \overline{W}$ and $x-y \in B_0(2n)$. Thus $x-y \in D$, so that $\{tx + (1-t)y | t \in [0,1]\} \subset A$. Choose semi-norms λ on X_1 and ν on X_2 such that $\|Dg_w(v,z)\| \leq \lambda(v)\|z\| + \|v\|\nu(z)$ for all $w \in A$, $v \in X_1$, $z \in X_2$. We may assume $\lambda \geq \mu$. Then, if $x,y \in C$ with $\lambda(x-y) \leq 1$,

$$\begin{aligned}\|(g(y)-g(x))(z)\| &\leq \int_0^1 \|Dg_{ty+(1-t)x}(y-x,z)\|dt \\ &\leq \lambda(y-x)\|z\| + \|y-x\|\nu(z)\end{aligned}$$

Since C is compact, there exists $m \in N$ such that $\sup\{\|g(y)-g(x)\| \mid x,y \in C\} \leq m$. Thus, if $x,y \in C$ with $\lambda(x-y) \geq 1$, $\|(g(y)-g(x))(z)\| \leq m\cdot\lambda(y-x)\|z\|$. Let $\beta = m\lambda$. Then $\|(g(y)-g(x))(z)\| \leq \beta(y-x)\|z\| + \|y-x\|\nu(z)$ for all $x,y \in C$, $z \in X_2$. $\square$

<u>5.18 Example</u>. Let X_1, X_2, X_3 be bw* spaces, U an open subset of X_1, and $g: U \to L(X_2,X_3)$ a C^1 map. Assume in addition that at least one of the three spaces is finite-dimensional. Then g is $\mathfrak{U}$-regulated.

<u>Proof</u>. Let C be a compact subset of U. Then $Dg(C)$ is compact in $L^2(X_1,X_2;X_3)$, and hence is bounded in $L^2(E_1,E_2;E_3)$. Thus there exists an integer $m \in N$ such that $\|Dg_x(v,z)\| \leq m\|v\|\cdot\|z\|$ for all $x \in C$, $v \in X_1$, $z \in X_2$. If X_2 is finite-dimensional, then the norm on X_2 is a weakly continuous semi-norm, so that 5.25 implies that g is $\mathfrak{U}$-regulated. The same reasoning applies if X_1 is finite-dimensional. So assume that X_3 is finite-dimensional. Note that this implies that X_3 is a normed space. Let C be a compact subset of U. By 5.7, there exists a neighborhood W of C, and semi-norms λ on X_1 and ν on X_2, such that $Dg(W) \subset L^2((X_1)_\lambda,(X_2)_\nu;X_3)$ and such that the map $Dg : W \to L^2((X_1)_\lambda,(X_2)_\nu;X_3)$ is continuous. Since C is compact,

there exists $n \in N$ such that $\sup\{\|Dg_x\|_{\tilde{\lambda},\tilde{\nu}} | x \in C\} \le n$. Thus $\|Dg_x(v,z)\| \le n\lambda(v)\nu(z)$ for all $x \in C$, $v \in X_1$, $z \in X_2$. Now, there exists $m \in N$ such that $\nu(z) \le m\|z\|$ for all $z \in X_2$, so that $\|Dg_x(v,z)\| \le nm\lambda(v)\|z\|$. Thus g is $\mathcal{U}$-regulated. □

5.19 Definition. Let X_1, X_2 be bw* spaces, U an open subset of X_1, $f : U \to X_2$. We will call f a (C^1) $\mathcal{U}$-map if f is C^1 and Df is $\mathcal{U}$-regulated. We will say that f is a C^k $\mathcal{U}$-map, $k > 1$, if f is a $\mathcal{U}$-map and f is C^k.

5.20 Lemma. Let X be a bw* space. Then scalar multiplication $m : R \times X \to X$, and addition, $add: X \times X \to X$, are both smooth $\mathcal{U}$-maps.

Proof. Since addition is a linear map from $X \times X$ to X, $D^2(add) = 0$, and hence add is a $\mathcal{U}$-map by Proposition 5.17.

To see that scalar multiplication is a $\mathcal{U}$-map, note that $Dm(r,x)(t,z) = tx + rz$ for all (r,x), $(t,z) \in R \times X$. Thus, for $(r,x),(s,y),(t,z) \in R \times X$, $\|(Dm(r,x) - Dm(s,y))(t,z)\| \le |t|\cdot\|x-y\| + |r-s|\cdot\|z\|$. Since R is finite-dimensional, the norm on R is weakly continuous, which implies that m is a $\mathcal{U}$-map. □

5.21 Lemma. Let X_1, X, X_2, X_3 be bw* spaces, U an open subset of X_1, Y an open subset of X, $f : U \to Y$ a C^1 map, and $g : Y \to L(X_2,X_3)$ a $\mathcal{U}$-regulated map. Then $g \circ f$ is $\mathcal{U}$-regulated.

Proof. Let C be a compact subset of U. Then there exists a convex balanced neighborhood W of zero in X_1 such that $C + \overline{W} \subset U$. Since C is bounded in X_1, there exists $n \in N$ such that $C \subset B_0(n)$. Let $D = \overline{W} \cap B_0(2n)$, and let $A = C + D$. Since A is compact, $f(A)$ is compact. Thus there exist semi-norms μ on X and ν on X_2 such that $\|(g(u)-g(v))(z)\| \le \mu(u-v)\|z\| + \|u-v\|\nu(z)$ for all $u,v \in f(A)$, $z \in X_2$. Consider $(Df)_\mu : U \to L(X_1,X_\mu)$. Since A is compact, there exists a semi-norm λ on X_1 such that $(Df)_\mu(A) \subset L_\lambda(X_1,X_\mu)$ and such that the map $(Df)_\mu : A \to L_\lambda(X_1,X_\mu)$ is continuous. We may assume $\{x \mid \lambda(x) < 1\} \subset W$, and we may assume $\|(Df)_\mu(x)\|_{\tilde{\lambda}} \leq 1$ for all $x \in A$. Now, if $x,y \in C$ and $\lambda(y-x) \leq 1$, then $y-x \in B_0(2n)$, so that $\{ty + (1-t)x \mid t \in [0,1]\} \subset A$. Thus

$$\mu(f(y)-f(x)) \leq \int_0^1 \|(Df)_\mu(ty + (1-t)x)\|_{\tilde{\lambda}}\, \lambda(y-x)\,dt$$
$$\leq \lambda(y-x).$$

Since A is compact, there exists $m \in N$ such that $\|Df(x)\| \leq m$ for all $x \in A$, which implies that

$$\|f(y)-f(x)\| \leq \int_0^1 \|Df(ty + (1-t)x)\| \cdot \|y-x\|\,dt \leq m\|y-x\|.$$

Thus, for $x,y \in C$ with $\lambda(y-x) \leq 1$,

$$\|((g\circ f)(y)-(g\circ f)(x))(z)\| \leq \lambda(y-x)\|z\| + m\|y-x\|\nu(z),$$

for all $z \in X_2$. Since C is compact, there exists $r \in N$ such that $\sup\{\|g\circ f(y)-g\circ f(x)\| \mid x,y \in C\} \leq r$. Thus, if $x,y \in C$ and $\lambda(y-x) > 1$, then $\|g\circ f(y)-g\circ f(x)\| \leq r\lambda(y-x)$, and we conclude that

$$\|(g\circ f(y)-g\circ f(x))(z)\| \leq r\lambda(y-x)\|z\| + m\|y-x\|\nu(z)$$

for all $x,y \in C$, $z \in X_2$. $\sqcap$

<u>5.22 Proposition</u>. Let X_1,X_2,X_3 be bw* spaces, U open in X_1, W open in X_2, $f : U \to W$, $g : W \to X_3$.

(a) If f and g are $\mathfrak{U}$-maps, then $g\circ f$ is a $\mathfrak{U}$-map.

(b) If $f : U \xrightarrow{\approx} W$ is a C^1 isomorphism and f is a $\mathfrak{U}$-map, then f^{-1} is a $\mathfrak{U}$-map.

(c) If $f \in L(X_1,X_2)$, then f is a $\mathfrak{U}$-map.

(d) If f is C^2 and X_2 is finite-dimensional, then f is a $\mathfrak{U}$-map.

(e) If f is C^2 and X_1 is finite-dimensional, then f is a $\mathfrak{U}$-map.

(f) If f is a $\mathfrak{U}$-map, and $h : U \to R$ is a C^2 function, then $h\cdot f$ is a $\mathfrak{U}$-map.

Proof. (a): $D(g\circ f) = ((Dg)\circ f)\cdot Df$. $Dg\circ f : U \to L(X_2,X_3)$ is $\mathcal{U}$-regulated by 5.21, so 5.10 implies that $D(g\circ f)$ is $\mathcal{U}$-regulated. (b): $D(f^{-1}) = (Df)^{-1}\circ f^{-1}$. Now, 5.11 implies that $(Df)^{-1}$ is $\mathcal{U}$-regulated, so another application of 5.21 tells us that $D(f^{-1})$ is $\mathcal{U}$-regulated. (c): $D^2 f = 0$, so 5.17 implies that Df is $\mathcal{U}$-regulated. (d) and (e) both follow from 5.18. Finally, to see (f), note that we have the following factorization of $h\cdot f$:

$U \xrightarrow{(h,f)} R \times X_2 \xrightarrow{m} X_2$. Since both factors are $\mathcal{U}$-maps, the composition is a $\mathcal{U}$-map by part (a). □

5.23 Proposition. Let X_1, X_2 be bw* spaces, U open in X_1, and $f : U \to X_2$ a set mapping. Assume that, for each $x \in U$, there exists a neighborhood U_x of x in U such that $f|_{U_x}$ is a $\mathcal{U}$-map. Then f is a $\mathcal{U}$-map.

Proof. Let C be a compact subset of U. For each $x \in C$, let W_x be a σ-compact convex open neighborhood of x such that $\overline{W_x} \subset U_x$, where $\overline{W_x}$ is the closure of W_x in X_1. Then $\{W_x \mid x \in C\}$ is an open cover of C. Since C is compact, there exists a finite subcover $W_1,\ldots,W_n$. For each i, let $\rho_i : X_1 \to R$ be a smooth non-negative function such that $\rho_i^{-1}((0,\infty)) = W_i$. Define $f_i : X_1 \to X_2$ by

$$f_i(x) = \begin{cases} \rho_i(x) f(x), & \text{if } x \in U \\ 0, & \text{if } x \in X_1 - W_i. \end{cases}$$

Then f_i is obviously C^1. Furthermore, 5.23 (f) implies that $f_i|_{U_i}$ is a $\mathcal{U}$-map. We will complete the proof by showing that $f_i : X_1 \to X_2$ is a $\mathcal{U}$-map for each $1 \le i \le n$. For, assuming this to be the case, let $W = \bigcup_{i=1}^{n} W_i$, and let $g = \Sigma_{i=1}^{n} f_i$. Then $g : X_1 \to X_2$ is a $\mathcal{U}$-map, so that $g|_W$ is a $\mathcal{U}$-map. Let $\rho = \Sigma_{i=1}^{n} \rho_i$, and define $\alpha : W \xrightarrow{C^\infty} R$ by $\alpha = 1/\rho$. Another application of 5.23 (f) yields that $\alpha\cdot(g|_W)$ is a $\mathcal{U}$-map. But

$\alpha\cdot(g|_W) = f|_W$. Thus there exists a semi-norm λ on X_1 such that, for $x,y \in C$, $z \in X_1$,

$\|(Df(y)-Df(x))(z)\| \le \lambda(y-x)\|z\| + \|y-x\|\lambda(z)$. Since this is true for each compact subset $C \subset U$, $f : U \to X_2$ is a $\mathcal{U}$-map. To see that $f_i : X_1 \to X_2$ is a $\mathcal{U}$-map, let A be a compact convex subset of X_1, and let $D = A \cap \overline{W_i}$. Then D is a compact subset of U_i. Since $f_i|_{U_i}$ is a $\mathcal{U}$-map, there exists a semi-norm ν on X_1 such that $\|(Df_i(y)-Df_i(x))(z)\| \le \nu(y-x)\|z\| + \|y-x\|\nu(z)$ for all $x,y \in D$, $z \in X_1$. I claim this inequality holds **also** for all $x,y \in A$, $z \in X_1$. For, it is certainly true if $x,y \in A-D$, for then $Df_i(y) = Df_i(x) = 0$. So the only case we must check is $x \in D$, $y \in A-D$. Let $r = \sup\{t \in [0,1] \,|\, ty + (1-t)x \in W_i\}$, and let $w = ry + (1-r)x$. Then $w \in D$, so that

$$\|(Df_i(x)-Df_i(w))(z)\| \le \nu(w-x)\|z\| + \|w-x\|\nu(z)$$

$$\le \nu(y-x)\|z\| + \|y-x\|\nu(z).$$

But $w \in X_1 - W_i$, so that $Df_i(w) = Df_i(y) = 0$, i.e. $\|(Df_i(x)-Df_i(y))(z)\| = \|(Df_i(x)-Df_i(w))(z)\|$, and so we conclude that $f_i : X_1 \to X_2$ is a $\mathcal{U}$-map. $\sqcap$

We are now in a position to prove an inverse function theorem for $\mathcal{U}$-maps. The proof will require four lemmas:

<u>5.24 Lemma</u>. Let X be a bw* space, U an open subset of X, $f : U \to X$ a $\mathcal{U}$-map such that $Df(U) \subset \mathrm{Aut}(X)$. Let C be a compact convex subset of U, and $x_o \in C$. Then there exists a neighborhood Y of x_o in C such that $f|_{\overline{Y}}$ is injective.

<u>Proof</u>. Since C is compact, there exists an integer $n \in N$ such that $\sup\{\|(Df)^{-1}(x)\| \,|\, x \in C\} \le n/2$. Choose a semi-norm λ on X such that $\|(Df(y)-Df(x))(z)\| \le \lambda(y-x)\|z\| + \|y-x\|\lambda(z)$ for all $y,x \in C$, $z \in X$. Define $Y = \{y \in C \,|\, \lambda(y-x_o) < 1/n\}$. Let $x_1, x_2 \in \overline{Y}$. Then

$$f(x_2)-f(x_1) = Df_{x_1}(x_2-x_1) + \int_0^1 (Df_{tx_2+(1-t)x_1} - Df_{x_1})(x_2-x_1)\,dt,$$

so $\|f(x_2)-f(x_1)\| \geq \|Df_{x_1}(x_2-x_1)\| - \int_0^1 \|(Df_{tx_2+(1-t)x_1}-Df_{x_1})(x_2-x_1)\|dt.$

Now $\|Df_{x_1}(x_2-x_1)\| \geq \frac{2}{n}\|x_2-x_1\|$. Looking at the second term,

$$\|(Df_{tx_2+(1-t)x_1}-Df_{x_1})(x_2-x_1)\| \leq \lambda(t(x_2-x_1))\|x_2-x_1\|$$
$$+\|t(x_2-x_1)\|\cdot\lambda(x_2-x_1) = 2t\lambda(x_2-x_1)\cdot\|x_2-x_1\|$$

Thus, $\int_0^1 \|(Df_{tx_2+(1-t)x_1}-Df_{x_1})(x_2-x_1)\|dt \leq \int_0^1 \lambda(x_2-x_1)\|x_2-x_1\|2tdt$

$\leq \lambda(x_2-x_1)\cdot\|x_2-x_1\| \leq \frac{1}{n}\|x_2-x_1\|$, and we conclude that

$\|f(x_2)-f(x_1)\| \geq \frac{2}{n}\|x_2-x_1\| - \frac{1}{n}\|x_2-x_1\| = \frac{1}{n}\|x_2-x_1\|$. ◻

The next lemma was proved in [13] for metric spaces only, this case being sufficient for the proof of an inverse function theorem for $\mathcal{U}$-maps on Bw* spaces. I am indepted to J. P. Penot for his observation that the lemma is true for general topological spaces, and for the following proof:

<u>5.25 Lemma</u>. Let Y and Z be topological spaces, Z Hausdorff, $f : Y \to Z$ a continuous map, and C a compact subset of Y such that $f|_C$ is injective. Assume in addition that, for each $x \in Y$, there exists a neighborhood U_x of x such that $f|_{U_x}$ is injective. Then there exists an open neighborhood U of C such that $f|_C$ is injective.

<u>Proof</u>. Let Δ_Z denote the diagonal in $Z \times Z$, and consider the map $f \times f: Y \times Y \to Z \times Z$. Since $f \times f$ is continuous, $(f \times f)^{-1}(Z \times Z - \Delta_Z)$ is open in $Y \times Y$. Now let $W = (f \times f)^{-1}(Z \times Z - \Delta_Z) \cup \Delta_Y$. Then I claim that W is open in $Y \times Y$. For, if $(x,y) \in W$, and $x \neq y$, then $(f \times f)^{-1}(Z \times Z - \Delta_Z)$ is an open neighborhood of (x,y) which is contained in W. And for each $x \in Y$, $U_x \times U_x$ is an open neighborhood of (x,x) which is contained in W. Thus W is an open subset of $Y \times Y$. But, since $f|_C$ is injective, $C \times C \subset W$. Since C is compact, there exists an open subset U of Y such that

$C \times C \subset U \times U \subset W$. Finally, the fact that $U \times U$ is contained in W is equivalent to the desired conclusion, that $f|_U$ is injective. □

5.26 Lemma. Let X be a bw* space, U a convex open subset of X, $f : U \to X$ a $\mathcal{U}$-map such that $Df(U) \subset \mathrm{Aut}(X)$, and $x_o \in U$. Then there exists an open neighborhood W of x_o such that $W \subseteq U$ and $f|_W$ is injective.

Proof. By 2.39, there exists an increasing sequence $\{C_n\}$ of compact convex subsets of U such that $U = \varinjlim_{n \in N} C_n$. We may assume $x_o \in C_1$. We will use induction to show the existence of an increasing sequence $\{W_i\}$ of compact subsets of U such that:

(1) $W_i \subset C_i$

(2) W_{i+1} is a neighborhood of W_i in C_{i+1} for each $i \in N$

(3) $f|_{W_i}$ is injective for each $i \in N$.

Then, if we let $W = \bigcup_{i \in N} W_i$, $f|_W$ is obviously injective. Furthermore, since $W_i \cap C_n$ is a neighborhood of $W_{i-1} \cap C_n$ in C_n for each $i > n$, $W \cap C_n$ is open in C_n for each $n \in N$, which implies that W is an open subset of U. To see the existence of the sequence $\{W_i\}$, we proceed as follows: by 5.24, there exists a closed neighborhood W_1 of x_o in C_1 such that $f|_{W_1}$ is injective. Let $m > 1$, and assume we have W_i for $1 \leq i < m$. By Lemma 5.25, there exists an open neighborhood U_m of W_{m-1} in C_m such that $f|_{U_m}$ is injective. Since C_m is compact, and hence normal, there exists a closed neighborhood W_m of W_{m-1} in C_m such that $W_m \subset U_m$. □

5.27 Lemma. Let X be a bw* space, U an open subset of X and $f:U \to X$ an injective $\mathcal{U}$-map such that $Df(U) \subset \mathrm{Aut}(X)$. Assume that C is a compact subset of X, and y_0 is a point in C such that $y_0 \in f(U)$. Then there exists a closed neighborhood A of y_0 in C such that:

(1) $A \subset f(U)$

(2) $f^{-1}(A)$ is compact.

Proof. We may assume $y_0 = f^{-1}(y_0) = 0$, $Df_o = Id$, and that there exists a positive integer $n \in N$ such that $C = K(n)$. Let λ be a semi-norm on X such that $\lambda(z) \leq \|z\|$ for all $z \in X$, and such that there exists a positive number $\varepsilon \in (0,\frac{1}{2}]$ with $\lambda^{-1}([0,3\varepsilon]) \subset U$.

Let $D = K(n+1) \cap \lambda^{-1}([0,3\varepsilon])$. Then D is a compact subset of U. Since Df is Lipschitz on compact subsets of U, there exists a positive integer $k \in N$ such that
$\|Df(y) - Df(x)\| \leq k\|y-x\|$ for all $x,y \in D$.

Let $F = K(n) \cap \lambda^{-1}([0,\varepsilon])$. Then F is a compact subset of D, and $B_x(2\varepsilon) \subset D$ for each $x \in F$. Choose a positive integer $j \in N$ such that $\|(Df(x))^{-1}\| \leq j$ for all $x \in D$, and let $r = \min\{\varepsilon/j, 1/(4k(j)^2)\}$. Then an examination of the proof of the Banach space inverse function theorem for C^1 maps with uniformly Lipschitz first derivative reveals that $B_{f(x)}(r) \subset f(B_x(2\varepsilon))$ for each $x \in F$.

Now, since f is a $\mathcal{U}$-map, there exists a semi-norm ν on X such that

$$\|(Df(y) - Df(x))(z)\| \leq \|y-x\|\nu(z) + \nu(y-x)\cdot\|z\|$$

for all $x,y \in D$, $z \in X$. We may assume without loss of generality that $\nu \geq \lambda$.

Let $A = K(n) \cap \nu^{-1}([0,r/n])$. Then, if $x \in A$,

$$\|f(x) - x\| = \|\int_0^1 (Df_{tx} - Df_0)(x)\,dt\| \leq \int_0^1 \|(Df_{tx} - Df_0)(x)\|\,dt$$

$$\leq \int_0^1 (\|tx\|\cdot\nu(x) + \nu(tx)\cdot\|x\|)\,dt = \int_0^1 2t\|x\|\nu(x)\,dt = \|x\|\cdot\nu(x)$$

$$\leq n(r/n) = r .$$

Thus $x \in B_{f(x)}(r) \subset f(B_x(2\varepsilon))$. Furthermore, $f^{-1}(x) \in B_x(2\varepsilon) \subset D$, so $f^{-1}(A)$ is a closed subset of D, and hence is compact. □

5.28 Lemma. Let X be a bw* space, U an open subset of X, and $f : U \to X$ an injective $\mathcal{U}$-map such that $Df(U) \subset Aut(X)$. Then $f(U)$ is open in X, and $f : U \to f(U)$ is a homeomorphism.

Proof. Let C be a compact subset of X, and let $y_0 \in C \cap f(U)$. By the preceding lemma, there exists a neighborhood A of y_0 in C such that $A \subset f(U)$. It follows that $C \cap f(U)$ is an open subset of C. Since this is true for each compact subset C in X, and since X is compactly generated, we conclude that $f(U)$ is open in X.

To see that $f:U \to f(U)$ is a homeomorphism, let C be a compact subset of $f(U)$. For each $x \in C$, there exists a closed neighborhood A_x of x in C such that $f^{-1}(A_x)$ is compact. Since we can cover C by a finite number of such neighborhoods, it follows that $f^{-1}(C)$ is compact. Since $f:f^{-1}(C) \to C$ is continuous and $f^{-1}(C)$ is compact, $f|_{f^{-1}(C)}$ is a homeomorphism. Thus $f^{-1}|_C$ is continuous. Since this is true for each compact subset C in $f(U)$, and since $f(U)$ is compactly generated, it follows that f^{-1} is continuous. □

5.29 Inverse Function Theorem. Let X be a bw* space, U an open subset of X, and $f:U \to X$ a C^k $\mathcal{U}$-map. Let $x \in U$, and assume $Df(x) \in \mathrm{Aut}(X)$. Then there exists an open neighborhood W of x in U such that $f(W)$ is open in X, $f:W \to f(W)$ is bijective, and $f^{-1}:f(W) \to W$ is a C^k $\mathcal{U}$-map.

Proof. By 5.16, there exists a neighborhood Y of x in U such that $Df(Y) \subset \mathrm{Aut}(X)$. The preceding lemmas imply the existence of an open neighborhood W of x, $W \subseteq Y$, such that $f(W)$ is open in X and such that $f:W \to f(W)$ is a homeomorphism. Since a $\mathcal{U}$-map is strongly C^1, the inverse function theorem for Banach spaces implies that $f^{-1}:f(W) \to W$ is strongly C^1, and $D(f^{-1}) = (Df)^{-1} \circ f^{-1}$. Since $(Df)^{-1}$ and f^{-1} are both weakly continuous, $D(f^{-1})$ is weakly continuous. But a strongly C^1 map which is defined on a weakly open set, and which has a weakly continuous derivative, is weakly C^1. Finally f^{-1} is a $\mathcal{U}$-map by 5.22, and is C^k by 5.5. □

6. $\mathfrak{U}$-MAPS AND THE GEOMETRY OF FUNCTION SPACES

The reader should be aware by this point that we plan to propose Bw* manifolds as abstract models for manifolds of maps. Consider for a moment the tangent bundle of such a manifold: if M is a C^k Bw* manifold, then 4.14 implies that $T(M)$ is a C^{k-1} Bw* manifold. However, the tangent bundle of a C^k $\mathfrak{U}$-manifold will not in general be a $\mathfrak{U}$-manifold. To remedy this defect, we will introduce new differentiability classes for maps between bw* spaces.

<u>6.1.</u> Define the "tangent bundle" functor from the category of open subsets of bw* spaces and C^k maps between them to the category of open subsets of bw* spaces and C^{k-1} maps between them as follows: if X_1, X_2 are bw* spaces, U_i an open subset of X_i, and $f:U_1 \to U_2$ a C^k map, then $T(U_i) = U_i \times X_i$, and $Tf:U_1 \times X_1 \to U_2 \times X_2$ is defined by $Tf(x,z) = (f(x), Df(x)(z))$ for all $x \in U_1$, $z \in X_1$.

Note that, if $k > 1$ and $j \le k$, we have an induced functor T^j which sends C^k maps to C^{k-j} maps.

<u>6.2 Definition</u>. Let X_1, X_2 be bw* spaces, U an open subset of X_1, $k \in N \cup \{\infty\}$, and $f:U \to X_2$ a C^k map. We say that f is (of differentiability class) $\mathfrak{U}^k$ if $T^j f$ is a $\mathfrak{U}$-map for each $0 \le j < k$.

<u>6.3 Lemma</u>. Let $k > 1$. Then f is $\mathfrak{U}^k \Longleftrightarrow Tf$ is $\mathfrak{U}^{k-1}$.

<u>Proof</u>. If f is $\mathfrak{U}^k$, then the above definition implies that Tf is $\mathfrak{U}^{k-1}$. Conversely, assume that Tf is $\mathfrak{U}^{k-1}$. Since $k > 1$, Tf is a $\mathfrak{U}$-map. Note that $Tf(x,z) = (f(x), Df(x)(z))$ for each $x \in U$, $z \in X_1$. Since Tf is a $\mathfrak{U}$-map, it follows that each factor of Tf is a $\mathfrak{U}$-map, and in particular that the map $g : U \times X \to X$ which is defined by $g(x,z) = f(x)$ is a $\mathfrak{U}$-map. Clearly this

implies that f is a $\mathcal{U}$-map. Thus $T^0 f$ is a $\mathcal{U}$-map. Since Tf is $\mathcal{U}^{k-1}$, $T^j f$ is a $\mathcal{U}$-map for $1 \le j < k$. Thus $T^j f$ is a $\mathcal{U}$-map for $0 \le j < k$, which implies that f is $\mathcal{U}^k$. □

6.4 Proposition. Let X_1, X_2, X_3 be bw* spaces, U open in X_2, $f : U \to W$ and $g : W \to X_3$. Assume that f and g are $\mathcal{U}^k$. Then $g \circ f$ is $\mathcal{U}^k$.

Proof. Since f and g are both C^k, $g \circ f$ is C^k. Let $j < k$: since T^j is a functor, $T^j(g \circ f) = (T^j g) \circ (T^j f)$. But both $T^j g$ and $T^j f$ are $\mathcal{U}$-maps, which implies that $T^j(g \circ f)$ is a $\mathcal{U}$-map. □

6.5 Proposition. Let X be a bw* space, U and W open subsets of X, $f : U \xrightarrow{\approx} W$ a C^1 isomorphism. Assume in addition that f is $\mathcal{U}^k$. Then f^{-1} is $\mathcal{U}^k$.

Proof. 5.5 implies that f is a C^k isomorphism, so that f^{-1} is C^k. Since T^j is a functor, $T^j f : T^j U \to T^j W$ is a C^{k-j} isomorphism, and $(T^j f)^{-1} = T^j(f^{-1})$. Since $T^j(f^{-1})$ is C^1 and $T^j f$ is a $\mathcal{U}$-map, 5.22 implies that $T^j(f^{-1})$ is a $\mathcal{U}$-map. Thus f^{-1} is $\mathcal{U}^k$. □

6.6 Proposition. Let X_1, X_2 be bw* spaces, $f \in L(X_1, X_2)$. Then $f : X_1 \to X_2$ is $\mathcal{U}^\infty$.

Proof. We already know that linear maps are C^∞ and $\mathcal{U}^1$. Now, $Tf : X_1 \times X_1 \to X_2 \times X_2$ is given by $Tf(x,y) = (f(x), f(y))$, so that $Tf \in L(X_1 \times X_1, X_2 \times X_2)$. Since Tf is linear, it is $\mathcal{U}^1$, which implies that f is $\mathcal{U}^2$, i.e. all linear maps are $\mathcal{U}^2$. In particular, Tf is $\mathcal{U}^2$, which implies that f is $\mathcal{U}^3$. Proceeding by induction in this fashion, we conclude that f is $\mathcal{U}^\infty$. □

The preceding proposition is not true in general for multilinear maps. In fact, we will see in Chapter 10 that there exist bilinear maps which are not $\mathcal{U}$-maps. However, we do have the following:

<u>6.7 Proposition</u>. Let $X, X_1, \ldots, X_r$ be bw* spaces, $f \in L^r(X_1, \ldots, X_r; X)$. Assume in addition that f is $\mathcal{U}^1$. Then f is $\mathcal{U}^\infty$.

<u>Proof</u>. For each $1 \leq i \leq r$, define $g_i \in L^r(X_1 \times X_1, \ldots, X_r \times X_r; X)$ by $g_i(x_1, y_1, \ldots, x_r, y_r) = f(x_1, \ldots, x_{i-1}, y_i, x_{i+1}, \ldots, x_r)$, and define $\ell_i \in (X_1 \times X_1 \times \cdots \times X_r \times X_r, X_1 \times \cdots \times X_r)$ by $\ell_i(x_1, y_1, \ldots, x_r, y_r) = (x_1, \ldots, x_{i-1}, y_i, x_{i+1}, \ldots, x_r)$. Also, define $\ell \in L(X_1 \times \cdots \times X_r \times X_1 \times \cdots \times X_r, X_1 \times X_1 \times \cdots \times X_r \times X_r)$ by $\ell(x_1, \ldots, x_r, y_1, \ldots, y_r) = (x_1, y_1, \ldots, x_r, y_r)$, and $p \in L((X_1 \times \cdots \times X_r) \times (X_1 \times \cdots X_r), X_1 \times \cdots \times X_r)$ by projection onto the first factor. Note that

$$Df(x_1, \ldots, x_r)(y_1, \ldots, y_r) = \Sigma_{i=1}^r f(x_1, \ldots, x_{i-1}, y_i, x_{i+1}, \ldots, x_r),$$

which implies that $Tf = (f \circ p, \Sigma_{i=1}^r g_i \circ \ell)$. Now, $g_i = f \circ \ell_i$, which implies that each g_i is $\mathcal{U}^1$. Thus Tf is $\mathcal{U}^1$, which implies that f is $\mathcal{U}^2$, which in turn implies that each g_i is $\mathcal{U}^2$. By induction, f is $\mathcal{U}^\infty$. □

<u>6.8 Corollary</u>. Let X be a bw* space. Then scalar multiplication $m : \mathbb{R} \times X \to X$ and $add : X \times X \to X$ are both $\mathcal{U}^\infty$.

<u>Proof</u>. By 5.20, m and add are both $\mathcal{U}^1$, so the above proposition implies that they are $\mathcal{U}^\infty$. □

Recall the criterion of Proposition 5.17 for a C^2 map to be $\mathcal{U}^1$. This result is an exceedingly important one, since it is the way we will show that the coordinate transformations in Sobolev manifolds are $\mathcal{U}^1$. We will next develop the analogous criterion for a C^k map to be $\mathcal{U}^{k-1}$. We will need one lemma:

<u>6.9 Lemma</u>. Let X_1, X_2 be bw* spaces, U an open subset of X_1, $k \in N$, and $f : U \to X_2$ a C^k map. Then, for each integer r

such that $0 \leq r < k$,

$$D^r(Tf)_{(x,y)}(v_1,w_1,\ldots,v_r,w_r)$$
$$= (D^r f_x(v_1,\ldots,v_r), D^{r+1}f_x(v_1,\ldots,v_r,y)$$
$$+ \Sigma_{i=1}^r D^r f_x(v_1,\ldots,v_{i-1},w_i,v_{i+1},\ldots,v_r))$$

Proof. For $r = 0$, this is just the definition of $Tf(x,y)$. The result for general r follows by induction, by differentiating the formula for the case $r-1$. □

6.10 Proposition. Let X_1, X_2 be bw* spaces, U an open subset of X_1, $k \geq 2$, and $f : U \to X_2$ a C^k map. Assume that, for each compact subset $C \subset U$ and integer r such that $2 \leq r \leq k$, there exists a semi-norm λ on X_1 such that

$$\|D^r f_x(v_1,\ldots,v_r)\| \leq \Sigma_{i=1}^r \lambda(v_1) \ldots \lambda(v_{i-1})\|v_i\|\lambda(v_{i+1}) \ldots \lambda(v_r)$$

for each $x \in C$, $v_i \in X_1$. Then f is u^{k-1}.

Proof. If $k = 2$, this is Proposition 5.17. So let $k \geq 2$, and assume the proposition proved for all j such that $2 \leq j \leq k$. We will prove it for $j = k+1$: let r be an integer such that $2 \leq r \leq k$, and let A be a compact subset of $T(U)$. Assume that $X_i \times X_i$ is normed by $\|(z_1,z_2)\| = \|z_1\| + \|z_2\|$ for each pair $(z_1,z_2) \in X_i \times X_i$. Let $C = p_1(A)$, where $p_1 : X_1 \times X_1 \to X_1$ is projection onto the first factor. Choose a semi-norm λ on $X_1 \times X_1$ such that

$$\|D^s f_x(v_1,\ldots,v_s)\| \leq \Sigma_{i=1}^s \lambda(v_1) \ldots \lambda(v_{i-1})\|v_i\|\lambda(v_{i+1}) \ldots \lambda(v_s)$$

for each $x \in C$, $v_i \in X_1$, and $s = r, r+1$. Define a continuous semi-norm ν on $X_1 \times X_1$ by $\nu((v,w)) = \lambda(v) + \lambda(w)$ for each $(v,w) \in X_1 \times X_1$. Let $(x,y) \in A$, $(v_i,w_i) \in X_1 \times X_1$. By the preceding lemma,

$$D^r(Tf)_{(x,y)}(v_1,w_1,\ldots,v_r,w_r)$$
$$= (D^r f_x(v_1,\ldots,v_r), D^{r+1}f_x(v_1,\ldots,v_r,y)$$
$$+ \Sigma_{i=1}^r D^r f_x(v_1,\ldots,v_{i-1},w_i,v_{i+1},\ldots,v_r))$$

Now,

$$\|D^r f_x(v_1,\dots,v_r)\| \le \Sigma_{i=1}^r \lambda(v_1) \ \dots\ \lambda(v_{i-1})\|v_i\|\lambda(v_{i+1}) \ \dots\ \lambda(v_r)$$

$$\le \Sigma_{i=1}^r \nu((v_1,w_1)) \ \dots\ \nu((v_{i-1},w_{i-1}))\|(v_i,w_i)\|\nu((v_{i+1},w_{i+1} \ \dots\ \nu((v_r,w_r)).$$

Choose an integer $m \in N$ such that $\lambda(y) \le m$ and $\|y\| \le m$ for all $(x,y) \in A$. Then

$$\|D^{r+1} f_x(v_1,\dots,v_r,y)\| \le m\ \Sigma_{i=1}^r \lambda(v_1) \ \dots\ \lambda(v_{i-1})\|v_i\|\lambda(v_{i+1}) \ \dots\ \lambda(v_r)$$

$$\le m\ \Sigma_{i=1}^r \nu((v_1,w_1)) \ \dots\ \nu((v_{i-1},w_{i-1}))\|(v_i,w_i)\|\nu((v_{i+1},w_{i+1})) \ \dots\ \nu((v_r,w_r)).$$

Similarly,

$$\|\Sigma_{i=1}^r D^r f_x(v_1,\dots,v_{i-1},w_i,v_{i+1},\dots,v_r)\|$$

$$\le r\ \Sigma_{i=1}^r \nu((v_1,w_1)) \ \dots\ \nu((v_{i-1},w_{i-1}))\|(v_i,w_i)\|\nu((v_{i+1},w_{i+1})) \ \dots\ \nu((v_r,w_r)),$$

which implies that

$$\|D^r(Tf)_{(x,y)}(v_1,w_1,\dots,v_r,w_r)\|$$

$$\le (m+r+1)\Sigma_{i=1}^r \nu((v_1,w_1)) \ \dots\ \nu((v_{i-1},w_{i-1}))\|(v_i,w_i)\|\nu((v_{i+1},w_{i+1})) \ \dots\ \nu((v_r,w_r))$$

for each $(x,y) \in A$, $(v_i,w_i) \in X_1 \times X_1$. By our inductive assumption, this implies that Tf is $\mathcal{U}^{k-1}$, which implies that f is $\mathcal{U}^k$. The proposition now follows for all k by induction. □

6.11 Remark. The condition on the higher-order differentials of f which is used in the above proposition to imply that f is $\mathcal{U}^{k-1}$ is itself preserved under composition and inverse, as can easily be verified for the cases $k = 2$ and $k = 3$. However, the proof for general k is complicated, and does not at this time seem important enough to bother with.

6.12 Corollary. Let X_1, X_2 be bw* spaces, U an open subset of X_1, $k \geq 2$, and $f : U \to X_2$ a C^k map. Assume in addition that X_2 is finite-dimensional. Then f is $\mathcal{U}^{k-1}$.

Proof. Since X_2 is finite-dimensional, we may assume that it is a normed space. Let C be a compact subset of U, and r an integer such that $2 \leq r \leq k$. By the fundamental property for D-spaces, there exists a semi-norm λ on X_1 such that $D^r(f)(C) \subset L^r_\lambda(X_1, X_2)$ and such that $D^r f : C \to L^r_\lambda(X_1, X_2)$ is continuous. Choose $m \in N$ such that $\|D^r f(x)\|_{\tilde{\lambda}} \leq m$ for all $x \in C$, and $\lambda(v) \leq m\|v\|$ for all $v \in X_1$. Then

$$\|D^r f_x(v_1, \ldots, v_r)\| \leq m\lambda(v_1) \ldots \lambda(v_r) \leq m^2\|v_1\|\lambda(v_2) \ldots \lambda(v_r)$$

for all $x \in C$, $v_i \in X_1$, so that the preceding proposition implies that f is $\mathcal{U}^{k-1}$. □

We have not yet attempted to relate the notion of a $\mathcal{U}$-map to any of the classes of maps familiar in analysis. What we will now show is that a $\mathcal{U}$-map is a generalization of the notion of Lipschitz map to the category of bw* spaces.

6.13 Lemma. Let X_1, X_2 be bw* spaces, U an open subset of X_1, $g : U \to L(X_1, X_2)$ a $\mathcal{U}$-regulated map, and C a compact subset of U. Then $g : C \to L(X_1, X_2)$ is strongly Lipschitz.

Proof. Choose a semi-norm λ on X_1 such that $\|(g(y)-g(x))(z)\| \leq \lambda(y-x)\|z\| + \|y-x\|\lambda(z)$ for all $x, y \in C$, $z \in X_1$. Choose $m \in N$ such that $\lambda(z) \leq m\|z\|$ for all $z \in X_1$. Then $\|(g(y)-g(x))(z)\| \leq m\|y-x\| \cdot \|z\| + m\|y-x\| \cdot \|z\|$, which implies that $\|g(y)-g(x)\| \leq 2m\|y-x\|$ for all $x, y \in C$. □

6.14 Lemma. Let X_1, X_2 be bw* spaces, U open in X_1, $f : U \to X_2$ a $\mathcal{U}^k$ map. Then, for each integer j such that $1 \leq j \leq k$, $D^j f$ is Lipschitz on each compact subset of U.

Proof. We proceed by induction. The result for $k = 1$ follows from 6.13. So let $k > 1$, and assume the lemma proved for all integers j such that $1 \le j \le k$. Let f be $\mathcal{U}^k$. Then f is $\mathcal{U}^{k-1}$, so $D^j f$ is Lipschitz on each compact subset of U for each $1 \le j \le k-1$. Also, Tf is $\mathcal{U}^{k-1}$, so that Tf is Lipschitz on each compact subset of $T(U)$ for $1 \le j \le k-1$.

Let C be a compact subset of U, and let $A = C \times B_0(1)$, so that $A \subset T(U)$. Let $v_1,\dots,v_k \in X_1$ such that $\|v_i\| = 1$ for each $1 \le i \le k$, and let $w_i = v_i$. Choose $m \in N$ such that m is a Lipschitz constant for $(D^{k-1}(Tf))|_A$ and for $(D^{k-1}f)|_C$. Let $x,y \in C$. From the formula of 6.9, we have:

$$\|(D^k f_y - D^k f_x)(v_1,\dots,v_k)\|$$

$$\le \|(D^{k-1}(Tf)_{(y,v_k)} - D^{k-1}(Tf)_{(x,v_k)})(v_1,v_1,\dots,v_{k-1},v_{k-1})\|$$

$$+ \|(D^{k-1}f_y - D^{k-1}f_x)(v_1,\dots,v_{k-1})\| + \Sigma_{i=1}^r \|(D^{k-1}f_y - D^{k-1}f_x)(v_1,\dots,v_{k-1})\|,$$

which implies that

$$\|D^k f(y) - D^k f(x)\| \le m\|(y,v_k)-(x,v_k)\| + m\|y-x\| + rm\|y-x\|.$$

Since $\|(y,v_k)-(x,v_k)\| = \|y-x\|$, $\|D^k f(y) - D^k f(x)\| \le m(r+2)\|y-x\|$. □

<u>6.15 Proposition</u>. Let X_1, X_2 be bw* spaces, U open in X_1, $k > 1$, and $f : U \to X_2$ a $\mathcal{U}^{k-1}$ map. Then f is strongly C^{k-}.

<u>Proof</u>. Immediate from 6.14. □

<u>6.16 Proposition</u>. Let X_1, X_2 be bw* spaces, U open in X_1, $k > 1$, and $f : U \to X_2$. Assume in addition that X_1 is finite-dimensional. Then f is $\mathcal{U}^{k-1}$ <==> f is strongly C^{k-}.

<u>Proof</u>. We need only show that, if f is strongly C^{k-}, then f is $\mathcal{U}^{k-1}$. Assume $k = 2$, and let C be a compact subset of U. Then Df is Lipschitz on C, so there exists $m \in N$ such that $\|Df(y) - Df(x)\| \le m\|y-x\|$ for all $x,y \in C$. This implies that $\|(Df(y) - Df(x))(z)\| \le m\|y-x\|\cdot\|z\|$ for all $x,y \in C$, $z \in X_1$. Since X_1 is finite-dimensional, the norm on X_1 is weakly continuous, which implies that Df is $\mathcal{U}$-regulated. Thus f is $\mathcal{U}^1$. If $k > 2$,

we use 6.9 to prove that Tf is strongly $C^{(k-1)-}$. Assume the result has been proved for $(k-1)$. Then Tf is $\mathcal{U}^{k-2}$, and hence f is $\mathcal{U}^{k-1}$. The result now follows by induction. □

<u>6.17 Proposition</u>. Let X_1, X_2 be bw* spaces, U open in X_1, $f : U \to X_2$ a $\mathcal{U}^k$ map, $h : U \to \mathbb{R}$ a C^{k+1} function. Then $h \cdot f$ is $\mathcal{U}^k$.

<u>Proof</u>. This is essentially the same as part (f) of 5.22, using 6.12, 6.8 and 6.4 in place of 5.20 and 5.22(a). □

<u>6.18 Lemma</u>. Let X_1, X_2, X_3 be bw* spaces, X_2 a subspace of X_3. Let U be open in X_1. Assume $f : U \to X_3$ is $\mathcal{U}^k$, and $f(U) \subset X_2$. Then $f : U \to X_2$ is $\mathcal{U}^k$.

<u>Proof</u>. Since X_2 is complete, it is a closed subspace of X_3. Thus, as we observed in 3.38, $f : U \to X_2$ is C^k. The definition of a $\mathcal{U}$-map implies that $f : U \to X_2$ is $\mathcal{U}^1$. To see that f is $\mathcal{U}^k$, note that $T^j f : T^j(U) \to (X_3)^{2^j}$ is $\mathcal{U}^1$ for all $0 \leq j < k$, and that $T^j f(T^j(U)) \subset (X_2)^{2^j}$. Thus $T^j f : T^j(U) \to (X_2)^{2^j}$ is $\mathcal{U}^1$, which implies that f is $\mathcal{U}^k$. □

We now have the tools which we need to return to the study of function spaces. In Chapter 1 we defined the notion of a LTS (linear topological space) section functor $\mathcal{M}$ on a compact C^∞ manifold M and showed that, for each smooth fiber bundle E over M, $\mathcal{M}(E)$ is a topological manifold. We can now introduce the notion of a D section functor $\mathcal{M}$ on M, and show that $\mathcal{M}(E)$ is a differentiable manifold.

<u>6.19 Definition</u>. Let M be a compact C^∞ manifold. A D section functor $\mathcal{M}$ on M is a LTS section functor on $FVB(M)$ such that, for each vector bundle ξ over M, $\mathcal{M}(\xi)$ is a projective D-space.

Let ξ, η be objects of $FVB(M)$, and let $f \in Map(\xi, \eta)$. Then, for each $x \in M$, $f|_{\xi_x}$ is a C^∞ map from ξ_x to η_x. Thus, for each $x \in M$ and $r \in N$, $D^r(f|_{\xi_x})$ is a smooth map from ξ_x to $L^r(\xi_x, \eta_x)$. Define a map

$\delta^r f : \xi \to L^r(\xi, \eta)$ by $(\delta^r f)|_{\xi_x} = D^r(f|_{\xi_x})$. Then $\delta^r f$ is obviously C^∞. If we make the standard identification of $L(\xi, L^r(\xi,\eta))$ with $L^{r+1}(\xi,\eta)$, then since $D((D^r f)|_{\xi_x}) = D^{r+1}(f|_{\xi_x})$, it is clear that $\delta(\delta^r f) = \delta^{r+1}(f)$. Recall that we defined a canonical linear injection $\mathcal{M}(L^r(\xi,\eta)) \to L^r(\mathcal{M}(\xi), \mathcal{M}(\eta))$ in 1.12, and showed that is continuous. This implies the following result:

<u>6.20 Lemma.</u> Let $\mathcal{M}$ be a locally convex section functor on M, ξ, η objects of $FVB(M)$, and $g \in Map(\xi, L^r(\xi,\eta))$. Define $\tilde{\mathcal{M}}(g) : \mathcal{M}(\xi) \to L^r(\mathcal{M}(\xi), \mathcal{M}(\eta))$ by $(\tilde{\mathcal{M}}(g))(s) = (g \circ s)$ for each $s \in \mathcal{M}(\xi)$. Then $\tilde{\mathcal{M}}(g)$ is continuous.

<u>6.21 Lemma.</u> Let $\mathcal{M}$ be a D section functor on M, ξ, η objects of $FVB(M)$, and $f \in Map(\xi, \eta)$. Then $\mathcal{M}(f)$ is C^1, and $D(\mathcal{M}(f)) = \tilde{\mathcal{M}}(\delta f)$.

<u>Proof.</u> By 3.28, it suffices to show that, for each $s_1, s_2 \in \mathcal{M}(\xi)$,

$$\mathcal{M}(f \circ s_2) - \mathcal{M}(f \circ s_1) = \int_0^1 \tilde{\mathcal{M}}(\delta f)_{(ts_2 + (1-t)s_1)}(s_2 - s_1)\,dt.$$

Since each side of the above equality is a section over M, it suffices to show that

$$f(s_2(x)) - f(s_1(x)) = \int_0^1 \delta f(ts_2(x) + (1-t)s_1(x))(s_2(x) - s_1(x))\,dt$$

for each $x \in M$. Since $\delta f(w) = D(f|_{\xi_x})(w)$ for each $w \in \xi_x$, and since $f|_{\xi_x}$ is a C^1 map from ξ_x to η_x, the classical Mean Value Theorem implies that

$$(f|_{\xi_x})(z) - (f|_{\xi_x})(y) = \int_0^1 D(f|_{\xi_x})(tz + (1-t)y)(z-y)\,dt$$

for each $y, z \in \xi_x$, and so the lemma is proved. □

<u>6.22 Theorem.</u> Let $\mathcal{M}$ be a D section functor on M, ξ, η objects of $FVB(M)$, and $f \in Map(\xi, \eta)$. Then $\mathcal{M}(f)$ is C^∞, and

$D^r(\mathcal{M}(f)) = \widetilde{\mathcal{M}}(\delta^r f)$ for each $r \in N$.

Proof. We proceed by induction. The preceding lemma gives us the result for $r = 1$. So let $r \in N$, and assume that we know that $\mathcal{M}(f)$ is C^r, and that $D^j(\mathcal{M}(f)) = \widetilde{\mathcal{M}}(\delta^j f)$ for each $1 \leq j \leq r$. Another application of 6.21 to the map $\delta^r f \in \mathrm{Map}(\xi, L^r(\xi,\eta))$ yields that $\mathcal{M}(\delta^r f)$ is C^1, and that $D(\mathcal{M}(\delta^r f)) = \widetilde{\mathcal{M}}(\delta(\delta^r f)) : \mathcal{M}(\xi) \to L(\mathcal{M}(\xi), \mathcal{M}(L^r(\xi,\eta)))$. Now, the canonical injection $\mathcal{M}(L^r(\xi,\eta)) \to L^r(\mathcal{M}(\xi),\mathcal{M}(\eta))$ is continuous and linear, and hence is C^∞. Since $\widetilde{\mathcal{M}}(\delta^r f)$ has the factorization

$$\mathcal{M}(\xi) \xrightarrow{\mathcal{M}(\delta^r f)} \mathcal{M}(L^r(\xi,\eta)) \longrightarrow L^r(\mathcal{M}(\xi),\mathcal{M}(\eta)),$$

$\widetilde{\mathcal{M}}(\delta^r f)$ is C^1, which implies that $\mathcal{M}(f)$ is C^{r+1}. Since the second term of the above factorization is linear, $D(\widetilde{\mathcal{M}}(\delta^r f))$ has the factorization

$$\mathcal{M}(\xi) \xrightarrow{\widetilde{\mathcal{M}}(\delta(\delta^r f))} L(\mathcal{M}(\xi),\mathcal{M}(L^r(\xi,\eta))) \longrightarrow L(\mathcal{M}(\xi), L^r(\mathcal{M}(\xi),\mathcal{M}(\eta)))$$

$$\approx L^{r+1}(\mathcal{M}(\xi),\mathcal{M}(\eta)).$$

But this composition is clearly $\widetilde{\mathcal{M}}(\delta^{r+1} f)$. Thus $\mathcal{M}(f)$ is C^{r+1}, and $D^j(\mathcal{M}(f)) = \widetilde{\mathcal{M}}(\delta^j f)$ for $1 \leq j \leq r+1$. □

6.23 Lemma. Let $\mathcal{M}$ be a D section functor on $FVB(M)$, ξ and η bundles of dimension q over M, and $f \in \mathrm{Map}(\xi,\eta)$ a smooth embedding of ξ into η. Then $\mathcal{M}(f)$ is a diffeomorphism of $\mathcal{M}(\xi)$ onto an open subset of $\mathcal{M}(\eta)$.

Proof. This is exactly the same as Lemma 1.8, except that 6.22 implies that all of the maps involved in the proof are C^∞. □

6.24 Theorem. Let $\mathcal{M}$ be a D section functor on $FB(M)$. Then $\mathcal{M}$ is a functor from $FB(M)$ to the category of smooth D-manifolds and C^∞ maps.

Proof. In 1.9, we showed that a section functor on $FVB(M)$ extends naturally and uniquely to a section functor on $FB(M)$. If we use 6.22 in that proof instead of 1.8, the present theorem follows immediately. $\square$

6.25 Example. Let $n \in N$. Then C^∞ is an n-dimensional section functor from $FB(n)$ to the category of smooth Frechet Schwartz manifolds and C^∞ maps.

The preceding discussion of function spaces is not original. Indeed, 6.20 through 6.24 were lifted essentially verbatim from Chapters 11 and 13 of [33]. The only difference is that Palais proved these results for Banach section functors, while we observed that the proofs work for all D section functors. However, this seemingly trivial observation does have at least one important consequence even in the case of Banach section functors. For recall our construction in Chapter 1 which associated a Bw* section functor on FVB(M) to each compact Banach section functor on FVB(M) which preserves boundedness. It is natural to ask whether every Bw* section functor arises in this fashion.

6.26 Theorem. Let M be a compact manifold. Then there exists a natural one-to-one correspondence between the set of Bw* section functors on FVB(M) and the set of compact Banach section functors on FVB(M) which preserve boundedness.

Proof. Given a Bw* section functor ω on FVB(M), we associate a functor $\mathfrak{M}$ to ω which we define as follows: for each object ξ of FVB(M), $\mathfrak{M}(\xi) = (\omega(\xi))''$, i.e. the underlying set of $\mathfrak{M}(\xi)$ is the underlying set of $\omega(\xi)$, and the topology on $\mathfrak{M}(\xi)$ is the strong topology associated to the Bw* topology on $\omega(\xi)$. For objects ξ, η of FVB(M) and $f \in Map(\xi,\eta)$, $\mathfrak{M}(f) : \mathfrak{M}(\xi) \to \mathfrak{M}(\eta)$ is defined to be the same mapping of sets as $\omega(f)$. To see that $\mathfrak{M}(f)$ is continuous, note that $\omega(f)$ is weakly C^∞ by 6.22, which implies that is is also strongly C^∞, i.e. $\mathfrak{M}(f)$ is a C^∞ mapping between Banach spaces. Let A be a bounded subset of $\mathfrak{M}(\xi)$. Then the closure of A in $\omega(\xi)$, $\overline{A}$, is compact, which implies that $\omega(f)(\overline{A})$ is compact in $\omega(\eta)$, which implies that $\mathfrak{M}(f)(\overline{A})$ is bounded in

$\mathcal{M}(\eta)$, i.e. $\mathcal{M}$ preserves boundedness. Finally, let K be a compact barrel in $\omega(\xi)$. Since the natural inclusion $\omega(\xi) \xrightarrow{i} C^0(\xi)$ is continuous, $i(K)$ is compact in $C^0(\xi)$. But K generates a norm on $\mathcal{M}(\xi)$ in which the closed unit ball is K itself, so that $\mathcal{M}$ is a compact Banach section functor. □

6.27 Corollary. Let M be a compact manifold, and let $\mathcal{M}$ be a functor on $FVB(M)$ which assigns, to each object ξ of $FVB(M)$, a Banach space $\mathcal{M}(\xi)$ of continuous sections of ξ such that the inclusion $\mathcal{M}(\xi) \to C^0(\xi)$ is compact (not just completely continuous), and which, to each morphism $f \in \mathrm{Map}(\xi,\eta)$, assigns a set mapping (not necessarily continuous) $\mathcal{M}(f) : \mathcal{M}(\xi) \to \mathcal{M}(\eta)$, $s \to f\circ s$. Assume in addition that, for each morphism f, $\mathcal{M}(f)$ sends bounded subsets of $\mathcal{M}(\xi)$ to bounded subsets of $\mathcal{M}(\eta)$. Then $\mathcal{M}(f)$ is a weakly C^∞ mapping for each morphism f of $FVB(M)$.

6.28 Theorem. The construction preceding Theorem 1.23 induces a one-to-one correspondence between the set of n-dimensional Bw* section functors and the set of n-dimensional compact Banach section functors which preserve boundedness.

Proof. Immediate from 6.26. □

Our next step should obviously be to show that the coordinate transformations in Bw* function spaces are $\mathcal{U}$-maps. Unfortunately, this is not the case in general. It IS true however, if the section functor under consideration is one of Palais' so-called "derivative" section functors. Our discussion of this class of section functors will be brief - see Chapter 5 of [33] for a more complete treatment.

Let M be a compact manifold. Given a Banach section functor $\mathcal{M}$ on $FVB(M)$, we define (for each non-negative integer k) its "k th derivative" section functor $\mathcal{M}_k$ as follows: for each object ξ of $FVB(M)$, the underlying set of $\mathcal{M}_k(\xi)$ is $\{s \in C^k(\xi) \mid j_k(s) \in \mathcal{M}(J^k(\xi))\}$, where $J^k(\xi)$ is the bundle of k-jets of ξ, and $j_k : C^k(\xi) \to C^0(J^k(\xi))$ is the k-jet extension map. By definition of $\mathfrak{M}_k(\xi)$, we have a natural injection $j_k : \mathfrak{M}_k(\xi) \to \mathfrak{M}(J^k(\xi))$. We topologize $\mathfrak{M}_k(\xi)$ by requiring that j_k be a homeomorphism, so that $\mathfrak{M}_k(\xi)$ is a normable space. Note that $\mathfrak{M}_o(\xi) = \mathfrak{M}(\xi)$. We have the following commutative diagram:

$$\begin{array}{ccc} \mathfrak{M}_k(\xi) & \longrightarrow & C^k(\xi) \\ {\scriptstyle j_k}\downarrow & & \downarrow{\scriptstyle j_k} \\ \mathfrak{M}(J^k(\xi)) & \longrightarrow & C^o(J^k(\xi)) \end{array}$$

Since $j_k : C^k(\xi) \to C^o(J^k(\xi))$ is a topological embedding, the continuity of the map $\mathfrak{M}_k(\xi) \to \mathfrak{M}(J^k(\xi)) \to C^o(J^k(\xi))$ implies that the inclusion $\mathfrak{M}_k(\xi) \to C^k(\xi)$ is continuous.

If ξ and η are vector bundles over M, $f \in \mathrm{Map}(\xi,\eta)$, define the continuous map $\mathfrak{M}_k(f) : \mathfrak{M}_k(\xi) \to C^k(\eta)$ by $\mathfrak{M}_k(f)(s) = f \circ s$ for each $s \in \mathfrak{M}_k(\xi)$. Note that $j_k(f\circ s) = J^k(f)\circ(j_k(s)) = \mathfrak{M}(J^k(f))(j_k(s))$. Since $j_k(s) \in \mathfrak{M}(J^k(\xi))$, this implies that $j_k(f\circ s) \in \mathfrak{M}(J^k(\eta))$. Thus $\mathfrak{M}_k(f)$ is a linear map from $\mathfrak{M}_k(\xi)$ to $\mathfrak{M}_k(\eta)$, and the following diagram is commutative:

$$\begin{array}{ccc} \mathfrak{M}_k(\xi) & \xrightarrow{\mathfrak{M}_k(f)} & \mathfrak{M}_k(\eta) \\ {\scriptstyle j_k}\downarrow & & \downarrow{\scriptstyle j_k} \\ \mathfrak{M}(J^k(\xi)) & \xrightarrow{\mathfrak{M}(J^k(f))} & \mathfrak{M}(J^k(\eta)) \end{array}$$

Since $\mathfrak{M}(J^k(f))$ is continuous, and the vertical maps are topological embeddings, we conclude that $\mathfrak{M}_k(f)$ is continuous. Also, note that we have the commutative diagram

$$\begin{array}{ccc} \mathfrak{M}_k(\xi) & \xrightarrow{i_{\ell k}} & \mathfrak{M}_\ell(\xi) \\ \downarrow{\scriptstyle j_k} & & \downarrow{\scriptstyle j_\ell} \\ (J^k(\xi)) & \xrightarrow{\mathfrak{M}(p_{\ell k})} & \mathfrak{M}(J^\ell(\xi)) \end{array}$$

for $k \geq \ell$, where $i_{\ell k}$ is the inclusion map, and $p_{\ell k}$ is the projection of the k-jets of ξ onto the ℓ-jets of ξ. Thus $i_{\ell k}$ is continuous. In particular, since $\mathfrak{M}_o = \mathfrak{M}$, the inclusion

$\mathcal{M}_k(\xi) \to \mathcal{M}(\xi)$ is continuous.

6.29 Proposition. (a) $\mathcal{M}_k$ is a Banach section functor.

(b) If $\mathcal{M}$ is compact, then $\mathcal{M}_k$ is compact, and $j_k : \mathcal{M}_k(\xi) \to \mathcal{M}(J^k(\xi))$ is a closed embedding of the associated Bw* spaces.

(c) If $\mathcal{M}$ preserves boundedness, then $\mathcal{M}_k$ preserves boundedness.

Proof. (a). Consider the commutative diagram

$$\begin{array}{ccc} \mathcal{M}_k(\xi) & \longrightarrow & C^k(\xi) \\ \downarrow j_k & & \downarrow j_k \\ \mathcal{M}(J^k(\xi)) & \xrightarrow{\tilde{i}} & C^o(J^k(\xi)) \end{array}$$

Let s_n be a Cauchy sequence in $\mathcal{M}_k(\xi)$, so that s_n is also a Cauchy sequence in $C^k(\xi)$. Thus there exists $s \in C^k(\xi)$ such that $s = C^k\text{-}\lim_{n\to\infty} s_n$. Similarly, $j_k(s_n)$ is Cauchy in $\mathcal{M}(J^k(\xi))$, so there exists $t \in \mathcal{M}(J^k(\xi))$ such that $t = \mathcal{M}\text{-}\lim_{n\to\infty} j_k(s_n)$.

Since each map in the above diagram is an inclusion, $t = j_k(s)$, which implies that $s \in \mathcal{M}_k(\xi)$. Now, since j_k is a topological embedding and $j_k(s)$ is the limit of $\{j_k(s_n)\}_{n\in N}$ in $\mathcal{M}(J^k(\xi))$, we conclude that s is the limit of $\{s_n\}_{n\in N}$ in $\mathcal{M}_k(\xi)$. Thus $\mathcal{M}_k(\xi)$ is complete.

(b): Assume $\mathcal{M}$ is compact. Identify $C^k(\xi)$ with $j_k(C^k(\xi))$, and $\mathcal{M}_k(\xi)$ with $j_k(\mathcal{M}_k(\xi))$. Referring to the diagram in part (a), let X be the space $\mathcal{M}(J^k(\xi))$ with the Bw* topology which it inherits from $C^o(J^k(\xi))$. Then $\tilde{i} : X \to C^o(J^k(\xi))$ is continuous. Since $C^k(\xi)$ is a closed subspace of $C^o(J^k(\xi))$, and since $\mathcal{M}_k(\xi) = \tilde{i}^{-1}(C^k(\xi))$, $\mathcal{M}_k(\xi)$ is a weakly closed subspace of X. Let X_k be $\mathcal{M}_k(\xi)$ with the Bw* topology which it inherits as a subspace of X. Then the inclusion $X_k \to C^k(\xi)$ is continuous, which implies that the inclusion $X_k \to C^o(\xi)$ is continuous. But the continuity of this last inclusion implies that the inclusion $\mathcal{M}_k(\xi) \to C^o(\xi)$ is compact, and that the Bw* topology induced on $\mathcal{M}_k(\xi)$ by this inclusion is X_k. Finally, from the way we defined X_k, it is clear that $j_k : X_k \to X$ is a continuous embedding.

(c): Let ξ, η be objects of FVB(M), and let $f \in \mathrm{Map}(\xi,\eta)$. We have the following commutative diagram, where the vertical maps are topological embeddings:

$$\begin{array}{ccc} \mathcal{M}_k(\xi) & \xrightarrow{\ \mathcal{M}_k(f)\ } & \mathcal{M}_k(\eta) \\ \Big\downarrow j_k & & \Big\downarrow j_k \\ \mathcal{M}(J^k(\xi)) & \xrightarrow{\ \mathcal{M}(J^k(f))\ } & \mathcal{M}(J^k(\eta)) \end{array}$$

Since $\mathcal{M}(J^k(f))$ preserves boundedness, so does $\mathcal{M}_k(f)$. $\square$

There are certain additional results which we will need which we can only obtain by using local coordinates. So assume for the moment that D^n is the n-dimensional disc, and that $\mathcal{M}$ is a Banach section functor on $\mathrm{FVB}(D^n)$.

<u>6.30 Lemma.</u> $(\mathcal{M}_k)_\ell = \mathcal{M}_{k+\ell}$.

<u>Proof</u>. Since each bundle over D^n is trivializable, it will suffice to show that $(\mathcal{M}_k)_\ell(D^n,\mathbb{R}) = \mathcal{M}_{k+\ell}(D^n,\mathbb{R})$. Now, $\mathcal{M}_{k+\ell}(D^n,\mathbb{R})$ consists of all $C^{k+\ell}$ functions f on D^n such that $D^\alpha \in \mathcal{M}(D^n,\mathbb{R})$ for all $|\alpha| \le k+\ell$, normed by $\|f\|_{k+\ell} \le \sum_{|\alpha|\le k+\ell} \|D^\alpha f\|$. The other space, $(\mathcal{M}_k)_\ell(D^n,\mathbb{R})$, consists of all C^ℓ functions f on D^n such that $D^\beta f \in \mathcal{M}_k(D^n,\mathbb{R})$ for all $|\beta| \le \ell$, i.e. all $C^{k+\ell}$ functions f on D^n such that $D^{\lambda+\beta} f \in \mathcal{M}(D^n,\mathbb{R})$ for all $|\gamma| \le k$, $|\beta| \le \ell$, and is normed by $\|f\|_{k,\ell} \le \sum_{|\beta|\le\ell} \|D^\beta f\|_k$. Clearly these sets of functions are equal, and the two norms are equivalent. $\square$

<u>6.31 Definition</u>. We say that $\mathcal{M}$ satisfies the Rellich condition if the natural transformation $\mathcal{M}_1 \to \mathcal{M}$ is completely continuous.

6.32 Remark. Let $\mathcal{M}$ be compact and Rellich. Since $\mathcal{M}$ is compact, $\mathcal{M}_1 \to \mathcal{M} \to C^0$ is compact. But since the natural transformation $\mathcal{M}_1 \to \mathcal{M}$ is completely continuous, it too must be compact.

6.33 Lemma. Let $\mathcal{M}$ be compact and Rellich. Then $\mathcal{M}_k$ is compact and Rellich for each $k \in N$.

Proof. We already know that $\mathcal{M}_k$ is compact, so we must show that the natural transformation $(\mathcal{M}_k)_1 \to \mathcal{M}_k$ is completely continuous. Note that $(\mathcal{M}_k)_1 = \mathcal{M}_{k+1} = (\mathcal{M}_1)_k$. Let ξ be an object of $FVB(D^n)$. Then we have the following commutative diagram:

$$\begin{array}{ccc}
(\mathcal{M}_1)_k(\xi) & \longrightarrow & \mathcal{M}_k(\xi) \\
\Big\downarrow j_k & & \Big\downarrow j_k \\
\mathcal{M}_1(J^k(\xi)) & \longrightarrow & \mathcal{M}(J^k(\xi))
\end{array}$$

Since the vertical maps are embeddings and the lower horizontal inclusion is completely continuous, the upper horizontal inclusion is completely continuous. $\square$

In the next theorem, D_i will denote $\frac{\partial}{\partial x_i}$, $\|s\|$ will denote the norm of s in $\mathcal{M}(\xi)$ for each $s \in \mathcal{M}(\xi)$, and $\|s\|_1$ will denote the norm of s in $\mathcal{M}_1(\xi)$ for each $s \in \mathcal{M}_1(\xi)$.

6.34 Lemma. Assume $\mathcal{M}$ is a compact, Rellich, Banach section functor on $FVB(D^n)$ which preserves boundedness. Let ξ, η be objects of $FVB(D^n)$, $f \in Map(\xi,\eta)$, and $k > 0$. Then $\mathcal{M}_k(f) : \mathcal{M}_k(\xi) \to \mathcal{M}_k(\eta)$ is $\mathcal{U}^\infty$.

Proof. Note that $\mathcal{M}_{k+1} = (\mathcal{M}_k)_1$, so it suffices to assume that $k = 1$. Since ξ and η are trivializable, we may assume that both are trivial. Thus $\|s\|_1 = \|s\| + \Sigma_{i=1}^n \|D_i s\|$ for each $s \in \mathcal{M}(\xi), \mathcal{M}(\eta)$. We will apply the criterion of Proposition 6.10 to show that $\mathcal{M}_1(f)$ is $\mathcal{U}^\infty$. So let A be a bounded subset of $\mathcal{M}_1(\xi)$, and let

$r \geq 2$. For $s \in A$, and $v_1,\ldots,v_r \in \mathcal{M}_1(\xi)$, $D^r(\mathcal{M}_1(f))(s)(v_1,\ldots,v_r) = (\delta^r f\circ s)(v_1,\ldots,v_r)$, where $\delta^r f : \xi \to L^r(\xi,\eta)$. Thus

$$\|(\delta^r f\circ s)(v_1,\ldots,v_r)\|_1 = \|(\delta^r f\circ s)(v_1,\ldots,v_r)\| + \Sigma_{i=1}^n \|D_i((\delta^r f\circ s)(v_1,\ldots,v_r))\|$$

$$\leq \|(\delta^r f)\circ s\|\|v_1\| \cdots \|v_r\| + \Sigma_{i=1}^n \|D_i(\delta^r f\circ s)\|\|v_1\| \cdots \|v_r\|$$

$$+ \Sigma_{i=1}^n \Sigma_{j=1}^r \|\delta^r f\circ s\|\|v_1\| \cdots \|v_{j-1}\|\|D_i v_j\|\|v_{j+1}\| \cdots \|v_r\|$$

$$\leq \|\delta^r f\circ s\|_1\|v_1\| \cdots \|v_r\|$$

$$+ \Sigma_{j=1}^r \|\delta^r f\circ s\|_1\|v_1\| \cdots \|v_{j-1}\|\|v_j\|_1\|v_{j+1}\| \cdots \|v_r\|$$

Since $\mathcal{M}$ preserves boundedness, we may choose $m \in N$ so that $\sup\{\|\delta^r f\circ s\|_1 \mid s \in A\} \leq m$. Thus

$$\|(\delta^r f\circ s)(v_1,\ldots,v_r)\|_1$$

$$\leq m(2n+1)\, \Sigma_{j=1}^r \|v_1\| \cdots \|v_{j-1}\|\|v_j\|_1\|v_{j+1}\| \cdots \|v_r\|$$

But 6.32 implies that $\| \ \|$ is weakly continuous on $\mathcal{M}_1(\xi)$, and so we are finished. □

Now let $\mathcal{M}$ be an n-dimensional Banach section functor, so that $\mathcal{M}_k$ is a section functor on FVB(n) for each $k \in N$. It would be ideal to be able to show that each $\mathcal{M}_k$ is also an n-dimensional section functor, but I have not been able to determine whether or not this is true. However, recall that the main purpose in defining the concept of an n-dimensional section functor was to prove Lemma 1.19. We CAN show that the conclusion of Lemma 1.19 is valid for each $\mathcal{M}_k$, even if we don't know that $\mathcal{M}_k$ is an n-dimensional section functor.

6.35 Lemma. Let M be a compact n-dimensional manifold, $N_1,\ldots,N_r$ compact n-dimensional submanifolds whose interiors cover M, and ξ a vector bundle over M.

Define

$$\widetilde{\mathcal{M}}_k(\xi) = \{(s_1,\ldots,s_r) \in \bigoplus_{i=1}^{r} \mathcal{M}_k(\xi|_{N_i}) \mid s_i|_{N_j} = s_j|_{N_i}\},$$

and let $\Phi : \mathcal{M}_k(\xi) \to \widetilde{\mathcal{M}}_k(\xi)$

$$s \to (s|_{N_1},\ldots,s|_{N_r}).$$

Then Φ is an isomorphism of Banach spaces.

<u>Proof</u>. Note that $(J^k\xi)|_{N_i} = J^k(\xi|_{N_i})$, so that we have the following commutative diagram, where the horizontal maps are the restriction maps:

$$\begin{array}{ccc} \mathcal{M}_k(\xi) & \longrightarrow & \mathcal{M}_k(\xi|_{N_i}) \\ \downarrow j_k & & \downarrow j_k \\ \mathcal{M}(J^k(\xi)) & \longrightarrow & \mathcal{M}((J^k\xi)|_{N_i}) \end{array}$$

Since the vertical maps are topological embeddings, and the lower horizontal map is continuous, the upper horizontal map must also be continuous. This implies that Φ is continuous.

The closed graph theorem will imply that Φ is an isomorphism if we show that Φ is surjective. So let $(t_1,\ldots,t_r) \in \widetilde{\mathcal{M}}_k(\xi)$. Then $(t_1,\ldots,t_r) \in \widetilde{C}^k(\xi)$, so there exists $t \in C^k(\xi)$ such that $t|_{N_i} = t_i$. Also, $j_k t_i \in \mathcal{M}(J^k(\xi|_{N_i})) = \mathcal{M}((J^k\xi)|_{N_i})$, and $(j_k t_i)|_{N_j} = (j_k t_j)|_{N_i}$. Thus there exists $u \in \mathcal{M}(J^k\xi)$ such that $u|_{N_i} = j_k t_i$. But now we have the following commutative diagram, where all maps are injective:

$$\begin{array}{ccc} \mathcal{M}_k(\xi) & \longrightarrow & C^k(\xi) \\ \downarrow j_k & & \downarrow j_k \\ \mathcal{M}(J^k\xi) & \longrightarrow & C^o(J^k\xi) \end{array}$$

Thus $j_k t = u$, which implies that $t \in \mathcal{M}_k(\xi)$, and hence that Φ is surjective. $\square$

6.36 Proposition. Assume in the preceding lemma that $\mathcal{M}$ is compact and preserves boundedness, and let ω_k be the Bw* section functor on FVB(n) associated to $\mathcal{M}_k$. Then $\Phi : \omega_k(\xi) \to \widetilde{\omega}_k(\xi)$ is an isomorphism of Bw* spaces.

Proof. Lemma 6.35 implies that Φ is a linear bijection. By the closed graph theorem for bw* spaces, it suffices to show that Φ is continuous. The weak continuity of Φ follows from the following commutative diagram:

$$\begin{array}{ccc} \mathcal{M}_k(\xi) & \longrightarrow & \widetilde{\mathcal{M}}_k(\xi) \\ \downarrow & & \downarrow \\ C^o(\xi) & \longrightarrow & \widetilde{C}^o(\xi) \end{array}$$

6.37 Theorem. Assume $\mathcal{M}$ is a compact Rellich n-dimensional Banach section functor which preserves boundedness. Let $k \in N$, and let ω_k be the Bw* section functor on FVB(n) associated to $\mathcal{M}_k$. Let M be a compact n-dimensional manifold, ξ and η objects of FVB(M), and $f \in \mathrm{Map}(\xi,\eta)$. Then $\omega_k(f) : \omega_k(\xi) \to \omega_k(\eta)$ is u^{∞}.

Proof. Choose subdiscs $D_1^n,\ldots,D_r^n$ of M whose interiors cover M_1. Let $\xi_i = \xi|_{D_i^n}$, $\eta_i = \eta|_{D_i^n}$, and $f_i = f|_{D_i^n} : \xi_i \to \eta_i$. We have the following commutative diagram, where the vertical maps are topological linear embeddings:

$$\begin{array}{ccc} \omega_k(f) : \omega_k(\xi) & \longrightarrow & \omega_k(\eta) \\ \downarrow & & \downarrow \\ \bigoplus_{i=1}^{r} \omega_k(f_i) : \bigoplus_{i=1}^{r} \omega_k(\xi_i) & \longrightarrow & \bigoplus_{i=1}^{r} \omega_k(\eta_i) \end{array}$$

By 6.34, $\bigoplus_{i=1}^{r} \omega_k(f_i)$ is u^{∞}, and so we may apply 6.18 to conclude that $\omega(f)$ is u^{∞}. □

All of the examples of derivative functors encountered in analysis are n-dimensional functors. Although we did not need the n-dimensionality of these functors to prove the above theorem, we will now present a simple criterion for a derivative functor to be n-dimensional.

6.38 Proposition. Let $\mathcal{M}$ be an n-dimensional Banach section functor, and $k \in N$. Suppose that, for each embedding $i : D^n \to D^n$, the induced map $i^* : \mathcal{M}_k(D^n, R) \to \mathcal{M}_k(D^n, R)$ is surjective. Then $\mathcal{M}_k$ is an n-dimensional Banach section functor.

Proof. Let $\mathcal{N}$ be the restriction of $\mathcal{M}_k$ to $FVB(D^n)$. By Theorem 1.20, $\mathcal{N}$ extends to an n-dimensional section functor on FVB(n). Let M be a compact n-dimensional manifold, and ξ an object of FVB(M). It suffices to show that $\mathcal{N}(\xi) = \mathcal{M}_k(\xi)$. Choose subdiscs $D_1^n, \ldots, D_r^n$ of M whose interiors cover M, and let $\xi_i = \xi|_{D_i^n}$. Then Lemma 1.19 implies that

$\Phi : \mathcal{N}(\xi) \approx \tilde{\mathcal{N}}(\xi) \subset \bigoplus_{i=1}^{r} \mathcal{N}(\xi_i)$. By 6.35, $\Phi : \mathcal{M}_k(\xi) \approx \tilde{\mathcal{M}}_k(\xi) \subset \bigoplus_{i=1}^{r} \tilde{\mathcal{M}}_k(\xi_i)$. But ξ_i is a bundle over D_i^n, which implies that $\mathcal{N}(\xi_i) = \mathcal{M}_k(\xi_i)$, and hence also that $\tilde{\mathcal{N}}(\xi) = \tilde{\mathcal{M}}_k(\xi)$. It is now immediate that $\mathcal{N}(\xi) = \mathcal{M}_k(\xi)$. □

The following are standard examples from analysis. It can be trivially verified that the first two are derivative functors, and the reader who is familiar with Sobolev spaces will easily see, at least for $k \in N$, that the third is also an n-dimensional derivative functor.

6.39 Examples. (1) Let $k \in N$, $k > 1$. Then $C^{k-} = (C^{1-})_{k-1}$.

(2) Let $k \in N$, $0 < \alpha < 1$. Then $C^{k+\alpha} = (C^{\alpha})_k$.

(3) Let $p > 1$, $k > n/p$, and $r \in N$. Then $L^p_{k+r} = (L^p_k)_r$.

Let us now reexamine the proofs of Lemma 6.34 and Theorem 6.37. In the lemma, we used the fact that $\mathcal{M}_1$ was a section functor, but we did not actually need $\mathcal{M}$ to be a section functor. The only properties of $\mathcal{M}$ which we actually needed were that the inclusion $\mathcal{M}_1(D^n, R) \to \mathcal{M}(D^n, R)$ was Rellich, that $\mathcal{M}(D^n, R)$ was a Banach algebra, and that the differential operators $\frac{\partial}{\partial x_i}$ induced continuous maps

$\mathcal{M}_1(D^n,\mathbb{R}) \to \mathcal{M}(D^n,\mathbb{R})$ for each $1 \le i \le n$. And as for Theorem 6.37, if we had known that $\mathcal{M}_1$ was n-dimensional, we could have dispensed with Lemma 6.35 and instead used Lemma 1.19. In fact, we could have weakened the hypothesis still further to prove the following technical result:

6.40 Theorem. Let ω be an n-dimensional Bw* section functor, and let $\mathcal{M}$ be the associated Banach section functor. Assume that there exists a Banach space E of distributions on D^n which contains $\omega(D^n,\mathbb{R})$ and which has the following properties:

(1) The inclusion $\omega(D^n,\mathbb{R}) \to E$ is continuous, and E is a module over $\omega(D^n,\mathbb{R})$.

(2) The differential operator $\frac{\partial}{\partial x_i}$ induces a continuous map $\mathcal{M}(D^n,\mathbb{R}) \to E$ for each $1 \le i \le n$.

(3) $\|s\|_{\mathcal{M}(D^n,\mathbb{R})} \le \|s\|_E + \Sigma_{i=1}^n \|\frac{\partial s}{\partial x_i}\|_E$ for each $s \in \mathcal{M}(D^n,\mathbb{R})$.

Then $\omega(f)$ is $\mathcal{U}^\infty$ for all morphisms f of FVB(n).

The above result, the proof of which is analogous to 6.37 and which we therefore omit, applies to a few section functors which are not covered by 6.37:

6.41 Examples. (1) C^{1-}, the Lipschitz functor.

(2) L_k^p, for $1 < p < \infty$ and $k > \max\{n/p, 1/2\}$.

Proof. (1) Let $f \in C^{1-}(D^n,\mathbb{R})$. Then $\frac{\partial}{\partial x_i}(f) \in L^\infty(D^n,\mathbb{R})$ for $1 \le i \le n$, and $\frac{\partial}{\partial x_i} : C^{1-}(D^n,\mathbb{R}) \to L^\infty(D^n,\mathbb{R})$ is continuous. Since $L^\infty(D^n,\mathbb{R})$ is a $C^0(D^n,\mathbb{R})$-module, it is also a module over $C^{1-}(D^n,\mathbb{R})$ with the weak topology.

(2) We will make free use of the results of Chapter 9 of [33] concerning Sobolev spaces. Although the proofs of these statements are only presented there for k integer-valued, these results in fact hold for all $k \in \mathbb{R}$.

To prove that L_k^p satisfies the conditions of 6.40, it suffices to show that $L_{k-1}^p(D^n,\mathbb{R})$ is a module over $L_{k-\epsilon}^p(D^n,\mathbb{R})$ for some $\epsilon > 0$. For, since the inclusion $L_k^p(D^n,\mathbb{R}) \to L_{k-\epsilon}^p(D^n,\mathbb{R})$ is compact, it will then follow that $L^p_{k-1}(D^n,\mathbb{R})$ is a module over $L^p_k(D^n,\mathbb{R})$ with the weak topology. To see that $L^p_{k-1}(D^n,\mathbb{R})$ is indeed a module over some $L^p_{k-\varepsilon}(D^n,\mathbb{R})$, we will need to treat two separate cases.

First assume that $k \geq 1$. Then $k-1 \geq 0$. Thus, if $0 < \varepsilon < \min\{1, k-\frac{n}{p}\}$, then $L^p_{k-1}(D^n,\mathbb{R})$ is a module over $L^p_{k-\varepsilon}(D^n,\mathbb{R})$ by 9.7 of [33].

Next, assume that $k \in (\frac{1}{2}, 1)$. Then $k-1 < 0$, and it is not quite so easy to see that multiplication extends to an action of some $L^p_{k-\varepsilon}(D^n,\mathbb{R})$ on $L^p_{k-1}(D^n,\mathbb{R})$. However, we will be able to show this as follows: define $q \in (1,\infty)$ by the equation $\frac{1}{q} + \frac{1}{p} = 1$, and let $0 < \varepsilon < \min\{k - \frac{n}{p}, k - \frac{1}{2}\}$. Also, note that $1-k \in (0, \frac{1}{2})$, and that $L^q_{1-k}(D^n,\mathbb{R})$ is the dual space of $L^p_{k-1}(D^n,\mathbb{R})$. Thus, if we can show that multiplication extends to a continuous action of $L^p_{k-\varepsilon}(D^n,\mathbb{R})$ on $L^q_{1-k}(D^n,\mathbb{R})$, then it will follow by taking the dual action that multiplication also extends to a continuous action of $L^p_{k-\varepsilon}(D^n,\mathbb{R})$ on $L^p_{k-1}(D^n,\mathbb{R})$.

Since $k - \frac{n}{p} > 0$, we have that $(1-k) - \frac{n}{q} = n - \frac{n}{q} - n + 1 - k = n(1 - \frac{1}{q}) - (n-1) - k = (\frac{n}{p} - k) - (n-1) < 0$. Thus part (3) of Theorem 9.5 of [33] immediately yields that $L^q_{1-k}(D^n,\mathbb{R})$ is a module over $L^p_{k-\varepsilon}(D^n,\mathbb{R})$. □

It is natural to ask whether $L^p_k(f)$ is $\mathcal{U}^\infty$ for morphisms f of FVB(n) if $n/p < 1/2$ and $n/p < k \leq 1/2$. I do not have any idea of what the answer to this question might be. Note that $L^p_{k-1}(D^n,\mathbb{R})$ is not a module over $L^p_{k-\varepsilon}(D^n,\mathbb{R})$ for any $\varepsilon > 0$, so that the "derivative functor" techniques of this chapter cannot be used to investigate this problem.

We will now digress for a moment to show that the Bw* manifolds we have been considering are both σ-compact and paracompact.

<u>6.42 Definition.</u> Let M be a compact manifold, E_1 an object of FB(M), and E_2 a closed C^∞ sub-bundle of E_1. Then a bundle tubular neighborhood of E_2 in E_1 is a smooth vector bundle U over E_2 such that the underlying manifold of U is an open neighborhood of E_2 in E_1, and such that the projection $r : U \to E_2$ is

a bundle morphism over M.

6.43 Lemma. Let M be a compact manifold, E_1 an object of $FB(M)$, and E_2 a closed C^∞ sub-bundle of E_1. Then there exists a bundle tubular neighborhood of E_2 in E_1.

Proof. This is analogous to the proof of the existence of open vector subbundles, and like the latter, may be found in Chapter 12 of [33]. □

6.44 Proposition. Let M be a compact manifold, and let $\mathfrak{M}$ be a LTS section functor on $FB(M)$. Suppose $\mathfrak{M}(M,R)$ is σ-compact (resp. paracompact). Then $\mathfrak{M}(E)$ is σ-compact (resp. paracompact) for each fiber bundle E over M.

Proof. Choose $m \in N$ sufficiently large so that there exists a closed fiber-preserving embedding $i : E \to R^m$. Let U be a bundle tubular neighborhood of E in R^m, $p : U \to E$ the projection map. Then $\mathfrak{M}(U)$ is an open submanifold of $\mathfrak{M}(M,R^m)$, and $\mathfrak{M}(p) : \mathfrak{M}(U) \to \mathfrak{M}(E)$ is continuous. This implies that $\mathfrak{M}(i) : \mathfrak{M}(E) \to \mathfrak{M}(M,R^n)$ is a homeomorphism of $\mathfrak{M}(E)$ onto its image, and it also implies that $\mathfrak{M}(i)(\mathfrak{M}(E)) = \{s \in \mathfrak{M}(M,R^n) \mid s(M) \subset i(E)\}$. Let A be the subspace of $\mathfrak{M}(M,R^n)$ whose underlying subset is $\{s \in \mathfrak{M}(M,R^n) \mid s(M) \subset i(E)\}$. Since the topology on $\mathfrak{M}(M,R^n)$ is stronger than the topology on $C^0(M,R^n)$, and since $i(E)$ is closed in R^n, we conclude that A is a closed subspace of $\mathfrak{M}(M,R^n)$. Thus A is a closed subspace of a σ-compact (resp. paracompact) space. □

6.45 Examples. Let M be a compact manifold, E an object of $FB(M)$, and ω a Bw* section functor on M.

(1) $\omega(E)$ is σ-compact and paracompact.

(2) $C^\infty(E)$ is paracompact.

6.46 Definition. Let M be a compact manifold. A $\mathfrak{u}$ section functor ω on $FVB(M)$ is a Bw* section functor such that $\omega(\xi)$ is a Bw* space for each object ξ of $FVB(M)$, and such that $\omega(f)$ is $\mathfrak{u}^\infty$ for each morphism f of $FVB(M)$.

6.47 Definition. Let $k \in N \cup \{\infty\}$. A $\mathcal{U}^k$ Bw* manifold X is a σ-compact Hausdorff space together with an atlas $\{(U_\lambda, \varphi_\lambda)\}_{\lambda \in \Lambda}$ on X such that:

(1) U_λ is open in X for each $\lambda \in \Lambda$, and $\bigcup_{\lambda \in \Lambda} U_\lambda = X$.

(2) $\varphi_\lambda : U_\lambda \to X_\lambda$ is a homeomorphism from U_λ onto an open subset of a Bw* space X_λ, for each $\lambda \in \Lambda$.

(3) $\varphi_\beta \circ \varphi_\alpha^{-1} : \varphi_\alpha(U_\alpha \cap U_\beta) \to \varphi_\beta(U_\alpha \cap U_\beta)$ is a $\mathcal{U}^k$ map for each pair $\alpha, \beta \in \Lambda$.

6.48 Theorem. Let M be a compact manifold, and assume ω is a $\mathcal{U}$ section functor on FVB(M). Then ω extends to a unique functor from FB(M) to the category of $\mathcal{U}^\infty$ Bw* manifolds and $\mathcal{U}^\infty$ maps.

Proof. Essentially the same as the proofs of Theorems 1.9 and 6.24. □

6.49 Examples. The following are n-dimensional $\mathcal{U}$ section functors:

(1) L^p_k, for $p > 1$ and $k > \max\{n/p, 1/2\}$.

(2) $C^{k+\alpha}$, for $k \in N$ and $0 < \alpha < 1$.

(3) C^{k-}, for $k \in N$.

This completes our discussion of function spaces. For the next two chapters we will return to the study of analysis in abstract Bw* spaces and manifolds. We will return to the discussion of function spaces when we consider smooth nonlinear differential operators and show that, if an operator induces a continuous map between two Sobolev spaces, then the map is often of differentiability class $\mathcal{U}^\infty$.

7. INFINITE-DIMENSIONAL DIFFERENTIABLE MANIFOLDS

The theory of differentiable maps between Bw* spaces will now be used to develop an abstract theory of differentiable Bw* manifolds. The basic results in differential topology which we will discuss are natural analogues of standard results about the differential topology of Banach manifolds; the material on Finsler structures and Riemannian metrics which will be presented was motivated by the investigations of K. Uhlenbeck into the calculus of variations.

This chapter must be regarded as more tentative in nature than the previous chapters. Abstract Bw* manifolds, as they will be defined here, much more closely resemble manifolds of maps than do Banach manifolds, but I suspect (for reasons which will be discussed in Chapter 10) that some of the analytical structure which manifolds of maps possess still remains to be isolated. Perhaps, in similar fashion to the way in which the concept of a $\mathcal{U}$-map had its origins in the calculus of variations, other additional structures which various manifolds of maps possess will be discovered through attempts to apply the emerging theory of Bw* manifolds to other concrete problems in partial differential equations.

<u>7.1 Definition.</u> Let $k \in N \cup \{\infty\}$. A C^k Bw* manifold M is a σ-compact Hausdorff space together with an atlas $\{(U_\lambda, \varphi_\lambda)\}_{\lambda \in \Lambda}$ on M such that:

(1) U_λ is open in M for each $\lambda \in \Lambda$, and $\bigcup_{\lambda \in \Lambda} U_\lambda = M$.

(2) $\varphi_\lambda : U_\lambda \to X_\lambda$ is a homeomorphism of U_λ onto an open subset of a Bw* space X_λ, for each $\lambda \in \Lambda$.

(3) $\varphi_\beta \circ \varphi_\alpha^{-1} : \varphi_\alpha(U_\alpha \cap U_\beta) \to \varphi_\beta(U_\alpha \cap U_\beta)$ is a C^k map for each $\alpha, \beta \in \Lambda$.

<u>7.2 Lemma.</u> M is paracompact.

Proof. Since M is σ-compact, there exists a countable collection of coordinate neighborhoods $\{U_n\}_{n\in N}$ which covers M. Each U_n is homeomorphic to an open subset of a Bw* space, so there exists a sequence of compact subsets C_{mn} of U_n such that the topology on U_n is coherent with the subspaces $\{C_{mn}\}_{m\in N}$. Since the U_n are open, the topology on M is coherent with $\{U_n\}_{n\in N}$, which implies that the topology on M is coherent with $\{C_{mn}\}_{m,n\in N}$, i.e. a subset $V \subseteq M$ is open in $M \Leftrightarrow V \cap C_{mn}$ is open in C_{mn} for each pair of integers $m,n \in N$. For each $r \in N$, define $A_r = \bigcup_{m,n\leq r} C_{mn}$. Then A_r is compact, and the topology on M is coherent with the topology on $\{A_r\}_{r\in N}$. But this means that $M = \varinjlim_{r\in N} A_r$. Now, it is standard point-set topology that a Hausdorff space which is the direct limit of an increasing sequence of compact subspaces is normal. Thus M is regular and σ-compact, which implies that M is paracompact. □

7.3 Definition. Let M_1, M_2 be C^k manifolds, $\{(U_\lambda, \varphi_\lambda) \mid \lambda \in \Lambda\}$ an atlas for M, $\{(V_\eta, \psi_\eta) \mid \eta \in H\}$ an atlas for M_2, $0 \leq \ell \leq k$, and $f : M_1 \to M_2$. We say that f is a C^ℓ map from M_1 to M_2 if $\psi_\eta \circ f \circ \varphi_\lambda^{-1}$ is C^ℓ for each $\eta \in H$, $\lambda \in \Lambda$.

7.4 Lemma. Let M be a C^k manifold, and U an open subset of M. Then there exists a C^k function $f : M \to R$ such that $f \geq 0$ and $f^{-1}((0,\infty)) = U$.

Proof. For each $x \in M$, there exists a chart (V_x, φ_x) with $x \in V_x$. Since $\{V_x \mid x\in M\}$ covers M, there exists a locally finite subcover $\{W_\lambda \mid \lambda\in\Lambda\}$. Let $\{U_\lambda \mid \lambda\in\Lambda\}$ be an open cover of M such that $\overline{U}_\lambda \subset W_\lambda$. For each $\lambda \in \Lambda$, let $(V_\lambda, \varphi_\lambda)$ be one of the original charts, chosen so that $W_\lambda \subset V_\lambda$. Since $\varphi_\lambda(U_\lambda)$ is an open subset of the Bw* space X_λ, there exists a C^k function $g_\lambda : X_\lambda \to R$, $g \geq 0$, such that $g^{-1}((0,\infty)) = \varphi_\lambda(U_\lambda \cap U)$. Define $f_\lambda : M \to R$ as follows:

$$f_\lambda|_{V_\lambda} = g_\lambda \circ \varphi_\lambda, \quad \text{and} \quad f_\lambda|_{M-V_\lambda} = 0 .$$

This definition immediately implies that f_λ is C^k on V_λ. Since $f_\lambda|_{M-\overline{U}_\lambda} = 0$, f_λ is also C^k on $M - \overline{U}_\lambda$. But the open sets V_λ and $M - \overline{U}_\lambda$ together cover M, so that f_λ is C^k on M. Finally, note that $\mathrm{supp}(f_\lambda) \subset \overline{U}_\lambda \subset W_\lambda$, so $\{\mathrm{supp}(f_\lambda) | \lambda \in \Lambda\}$ is locally finite. Thus, if we define $f : M \to \mathbb{R}$ by $f = \sum_{\lambda\in\Lambda} f_\lambda$, then f is C^k, $f \geq 0$, and $f^{-1}((0,\infty)) = U$. □

Recall that, in Chapter 2, (cf. the discussion preceding Example 2.31) we defined a functor J from the category of bw* spaces and continuous linear maps to the category of Banach spaces and continuous linear maps: if X was a Banach space, then $J(X) = E = X''$, and if $\ell \in L(X_1,X_2)$, then $J(\ell) \in L(E_1,E_2)$ was the linear map which induced the same map between the underlying sets as the linear map ℓ itself.

In Chapter 4 we showed that, if $k \geq 1$, we could extend J to a functor from the category of open subsets of bw* spaces and C^k maps to the category of open subsets of Banach spaces and C^{k-} maps.

Now we will see that J can be extended to a functor from the category of C^k Bw* manifolds and C^k maps to the category of C^{k-} Banach manifolds and C^{k-} maps. For, let M be a C^k Bw* manifold, and consider a defining system $\{U_\lambda,\varphi_\lambda\}_{\lambda\in\Lambda}$ of charts for the differentiable structure on M. Since $\varphi_\lambda(U_\lambda)$ is open in X_λ, it is open in $E_\lambda = X''_\lambda$. Also, since $\varphi_\eta \circ \varphi_\xi^{-1}: \varphi_\xi(U_\xi \cap U_\eta) \to \varphi_\eta(U_\xi \cap U_\eta)$ is weakly C^k, it is strongly C^{k-}. Thus, if we define a new topological space $J(M)$ with the same underlying set as M, by defining a subset $U \subseteq J(M)$ to be open $\Leftrightarrow$ $\varphi_\lambda(U_\lambda \cap U)$ is open in E_λ for each $\lambda \in \Lambda$, then the atlas on M automatically induces the structure of a C^{k-} Banach manifold on $J(M)$. Since the norm topology on each $\varphi(U_\lambda)$ is stronger than its Bw* topology, there exists a natural continuous map $J(M) \to M$. Thus $J(M)$ is Hausdorff. Finally, if M and N are C^k Bw* manifolds, and $f : M \to N$ is a C^k map, then $\psi_\eta \circ f \circ \varphi_\lambda^{-1}$ is strongly C^{k-} for each pair of charts ψ_η on N and φ_λ on M, which implies that we have a functor from the category of C^k Bw* manifolds to the category of C^{k-} Banach manifolds.

In particular, J is a functor from the category of smooth Bw* manifolds to the category of smooth Banach manifolds, and we see that the category of smooth Bw* manifolds is in reality a very special subcategory of the category of smooth Banach manifolds, with significantly more structure. The functor J can be regarded as a stripping functor which removes the additional structure.

<u>7.5 Theorem.</u> Assume $k \geq 2$, and let M be a C^k Bw* manifold. Then $J(M)$ admits a Finsler structure. If M is modeled on weak Hilbert space, then $J(M)$ admits a $C^{(k-1)-}$Riemannian metric.

<u>Proof</u>. We will show that, if M is a weak Hilbert manifold, then $J(M)$ admits a $C^{(k-1)-}$Riemannian metric (the proof of the assertion about Finsler structures is the same, if we replace the phrase "Riemannian metric" with "Finsler structure" throughout the discussion). Choose a set of charts $\{U_\lambda, \varphi_\lambda) \mid \lambda \in \Lambda\}$ whose domains form a locally finite cover of M. There exists a collection of open subsets $\{W_\lambda \mid \lambda \in \Lambda\}$ of M such that $\bigcup_{\lambda\in\Lambda} W_\lambda = M$ and such that $\overline{W}_\lambda \subset U_\lambda$ for each $\lambda \in \Lambda$. Let $f_\lambda : M \to \mathbb{R}$ be a C^k function such that $f_\lambda \geq 0$ and such that $f_\lambda^{-1}((0,\infty)) = W_\lambda$. Then $J(f_\lambda) : J(M) \to \mathbb{R}$ is strongly C^{k-}. Since $J(U_\lambda)$ inherits a $C^{(k-1)-}$Riemannian metric from Hilbert space via $J(\varphi_\lambda)$, we may use $\{J(f_\lambda) \mid \lambda\in\Lambda\}$ to patch together a globally-defined $C^{(k-1)-}$Riemannian metric on $J(M)$. □

<u>7.6 Corollary.</u> Let $k \geq 2$. Then $J(M)$ is paracompact.

<u>Proof.</u> It is a standard result of Banach manifold theory that, if F is a C^1 Banach manifold with a Finsler structure, then the Finsler structure may be used to induce a metric on F. If the underlying topology of F is regular, then the metric topology is identical to the original topology, which implies that F is paracompact. Thus, to show that $J(M)$ is paracompact, it suffices to show that $J(M)$ is regular.

So let $x \in M$, and let W be an open neighborhood of x in $J(M)$. Let (U,φ) be a chart on M such that $x \in U$, and let X be the Bw* space which contains $\varphi(U)$. Then $J(\varphi)(J(U) \cap W)$ is an open neighborhood of $\varphi(x)$ in $E = X''$, so there exists $\varepsilon > 0$ such that $B_{\varphi(x)}(\varepsilon) \subset J(\varphi)(J(U) \cap W)$. Now, $B_{\varphi(x)}(\varepsilon)$ is compact in X, which implies that $\varphi^{-1}(B_{\varphi(x)}(\varepsilon))$ is compact, and hence closed, in M. Since the map $J(M) \to M$ is continuous, $J(\varphi^{-1})(B_{\varphi(x)}(\varepsilon))$ is closed in $J(M)$. But x is an interior point of $J(\varphi^{-1})(B_{\varphi(x)}(\varepsilon))$ in $J(M)$. Since $J(\varphi^{-1})(B_{\varphi(x)}(\varepsilon)) \subset W$, this implies that $J(M)$ is regular. □

<u>7.7 Remark</u>. Douady has shown that every Hausdorff Banach manifold which admits a Finsler structure is a regular topological space, so that the above proof was not actually necessary. However, Douady's result is not widely known; and since the proof presented above makes use of the relation between the two topologies on a Bw* manifold, it seems worth while to include it here.

Our first result will be that the natural map $J(M) \to M$ is a homotopy equivalence. The proof of this theorem follows from the corresponding local result (which is a consequence of a theorem which Palais proved several years ago), and a theorem about maps between paracompact spaces due to J. Milnor, which is the device which enables us to globalize the local version. The specific theorems involved are the following:

<u>7.8 Theorem.</u> Let E and F be locally convex topological vector spaces, $\varphi : E \to F$ a continuous linear injection of E onto a dense linear subspace of F. Assume W is open in F, and let $U = \varphi^{-1}(W)$. Assume in addition that both U and W are paracompact. Then $\varphi : U \to W$ is a homotopy equivalence.

<u>Proof.</u> See Palais, [31]. □

<u>7.9 Theorem</u>. Let X and Y be paracompact topological spaces, $f : X \to Y$ a continuous map. Let $\{X_i\}_{i \in N}$ and $\{Y_i\}_{i \in N}$ be increasing sequences of open subsets of X and Y (respectively)

such that $X = \bigcup_{i\in N} X_i$ and $Y = \bigcup_{i\in N} Y_i$. Assume in addition that $f(X_i) \subset Y_i$ for each $i \in N$, and that $f : X_i \to Y_i$ is a homotopy equivalence. Then $f : X \to Y$ is a homotopy equivalence.

Proof. See the appendix to Milnor, [27]. □

The proof that the natural map $J(M) \to M$ is a homotopy equivalence will require one more result (Theorem 7.10), which I will also state without proof. The statement of 7.10 is similar to 7.9, and a very slight modification of Milnor's proof of 7.9 will yield a proof of 7.10.

7.10 Theorem. Let X and Y be normal spaces, and assume $X = X_1 \cup X_2$, $Y = Y_1 \cup Y_2$, X_i open in X and Y_i open in Y. Let $f : X \to Y$ be a continuous map such that $f(X_i) \subset Y_i$. Assume in addition that $f : X_i \to Y_i$ is a homotopy equivalence, and that $f : X_1 \cap X_2 \to Y_1 \cap Y_2$ is a homotopy equivalence. Then $f : X \to Y$ is a homotopy equivalence.

Proof. See paragraph which follows 7.9. □

7.11 Theorem. Let $k \geq 2$, and let M be a C^k Bw* manifold. Then the natural map $J(M) \to M$ is a homotopy equivalence.

Proof. Choose a countable cover $\{(U_i, \varphi_i)\}_{i\in N}$ of M by charts. For each $n \in N$, define $W_n = \bigcup_{i\leq n} U_i$. Then $M = \bigcup_{n\in N} W_n$, $J(M) = \bigcup_{n\in N} J(W_n)$, and $W_n \subseteq W_{n+1}$ for each $n \in N$. Let X_i be the Bw* space which contains $\varphi(U_i)$, and let $E_i = X_i''$. Then we have the following commutative diagram:

$$\begin{array}{ccc} J(U_i) & \longrightarrow & U_i \\ \cap\, J(\varphi_i) & & \cap\, \varphi_i \\ E_i & \longrightarrow & X_i \end{array}$$

Thus Theorem 7.8 implies that the natural map $J(U_i) \to U_i$ is a homotopy equivalence. We next show by induction that the map

$J(W_n) \to W_n$ is a homotopy equivalence for each $n \in N$. Note that $W_1 = U_1$, so that we already have this result for $n = 1$. So let $n > 1$, and assume that the map $J(W_{n-1}) \to W_{n-1}$ is a homotopy equivalence. Since $W_n = W_{n-1} \cup U_n$, and $J(W_n) = J(W_{n-1} \cup U_n) = J(W_{n+1}) \cup J(U_n)$, the result for W_n will follow from 7.10 if we can show that the map $J(W_{n-1}) \cap J(U_n) \to W_{n-1} \cap U_n$ is a homotopy equivalence. But $J(W_{n-1}) \cap J(U_n) = J(W_{n-1} \cap U_n)$, and $W_{n-1} \cap U_n$ is a subset of a coordinate neighborhood. Thus another application of 7.8 yields that the map $J(W_{n-1} \cap U_n) \to W_{m-1} \cap U_n$ is a homotopy equivalence. By induction, the map $J(W_n) \to W_n$ is a homotopy equivalence for each $n \in N$. And now, Theorem 7.9 implies that the map $J(M) \to M$ is a homotopy equivalence. □

We return now to the Finsler structure constructed in Theorem 7.5. That particular Finsler structure was constructed using a partition of unity on M rather than simply on $J(M)$, so that it does not seem unreasonable to expect that the Finsler structure will turn out to have properties which relate to the topology or geometry of M.

A very simple example of such a property is given in the next theorem. This property was first noticed by K. Uhlenbeck in connection with certain naturally-arising metrics which she constructed on manifolds of Sobolev maps.

<u>7.12 Theorem</u>. Let M be a Bw* manifold, and assume that $J(M)$ has the Finsler structure constructed in Theorem 7.5. Let (U, φ) be a chart on M, let X be the Bw* space which contains $\varphi(U)$, and let $E = X''$. Let $\| \quad \|$ be a norm on E, let $\| \quad \|_x$ be the norm on $T_x(J(\varphi(U)))$ induced by the Finsler structure for each $x \in \varphi(U)$, and let C be a compact subset of U. Then there exist constants $k_1, k_2 > 0$ such that $k_1\|v\| \le \|v\|_x \le k_2\|v\|$ for each $v \in X$, $x \in C$.

<u>Proof</u>. We will use the locally finite system of charts $\{U_\lambda : \lambda \in \Lambda\}$ and the functions $\{f_\lambda : \lambda \in \Lambda\}$ of 7.5.

Let $H = \{\lambda \in \Lambda | U_\lambda \cap C \neq \varphi\}$. Since C is compact, and the charts form a locally finite covering of M , it follows that H contains only a finite number of elements $\lambda_1, \ldots, \lambda_n$. To simplify the subscripts, we will write U_i for U_{λ_i} , f_i for f_{λ_i} , etc. Then, for each $x \in \varphi(C)$, $v \in X$, $\|v\|_x^2 = \Sigma_{i=1}^n (f_i \circ \varphi^{-1})(x) \|D(\varphi_i \circ \varphi^{-1})_x(v)\|^2$.

Since each $f_i \circ \varphi^{-1}$ is continuous on C , there is a positive integer N_0 such that $f_i(x) \leq N_0$ for all $x \in C$, $1 \leq i \leq n$. Now, $\varphi_i \circ \varphi^{-1}$ is not defined on all of $\varphi(C)$, but only on $\varphi(C \cap U_i)$. However, we need only worry about bounding the derivative of $\varphi_i \circ \varphi^{-1}$ on $\varphi(C \cap \overline{W}_i)$, where $\overline{W}_i = \text{supp}(f_i)$, and $\varphi(C \cap \overline{W}_i)$ is compact. Since $\varphi_i \circ \varphi^{-1}$ is a diffeomorphism, it follows that $D(\varphi_i \circ \varphi^{-1})$ and $(D(\varphi_i \circ \varphi^{-1}))^{-1}$ are both continuous maps from $\varphi(C \cap \overline{W}_i)$ to $L(X,X)$, and hence are both bounded. Thus there is a positive integer N_1 such that $\|D(\varphi_i \circ \varphi^{-1})(x)\| \leq N_1$ and $\|(D(\varphi_i \circ \varphi^{-1})(x))^{-1}\| \leq N_1$ for all $x \in \varphi(C \cap \overline{W}_i)$, for each $1 \leq i \leq n$. Finally, define $g : C \to \mathbb{R}$ by $g(x) = \max\{f_1 \circ \varphi^{-1}(x), \ldots, f_n \circ \varphi^{-1}(x)\}$. Then g is continuous on C and is positive at each point of C , so that it must be bounded away from zero, i.e. there exists $\varepsilon > 0$ such that $g(x) \geq \varepsilon$ for all $x \in C$.

It now follows immediately from the above that $\|v\|_x^2 \leq n \cdot N_0 \cdot N_1^2 \cdot \|v\|^2$ for each $x \in C$, $v \in X$, and also that $\|v\|_x^2 \geq \frac{\varepsilon}{N_1^2} \|v\|^2$. □

<u>7.13 Remark</u>. Assume that M is modeled on weak Hilbert space and that we have constructed a Riemannian metric in 7.5. It will be informative to examine this metric in the local chart U of the above theorem. For each $x \in \varphi(U)$, $v \in X$,

$$\begin{aligned} \langle v,v\rangle_x &= \Sigma_{\lambda\in\Lambda}(f_\lambda \circ \varphi^{-1})(x) \| D(\varphi_\lambda \circ \varphi^{-1})_x(v)\|^2 \\ &= \Sigma_{\lambda\in\Lambda}(f_\lambda \circ \varphi^{-1})(x) \langle (D(\varphi_\lambda \circ \varphi^{-1})_x)^* D(\varphi_\lambda \circ \varphi^{-1})_x v,v\rangle \end{aligned}$$

Thus, locally the metric is given by the map $A : \varphi(U) \to L(E,E)$ defined by

$$A(x) = \Sigma_{\lambda\in\Lambda}(f_\lambda\circ\varphi^{-1})(x)(D(\varphi_\lambda\circ\varphi^{-1})_x)^*D(\varphi_\lambda\circ\varphi^{-1})_x .$$

However, A is not in general a weakly continuous map from $\varphi(U)$ to $L(X,X)$. The problem here is basically that the adjoint map, while it is a continuous linear automorphism of $L(E,E)$, is not continuous when regarded as a linear automorphism of $L(X,X)$.

It is easy to verify directly that the adjoint operator is not continuous on $L(X,X)$ for X an infinite-dimensional weak Hilbert space. However, rather than show this, we will present an example of a metric on an open subset of Hilbert space which is defined as the pull-back of the usual metric via a nonlinear diffeomorphism, and show that the operator-valued function which represents this metric is not weakly continuous.

7.14 Example. Let S^1 be the unit circle, equipped with the Lebesque measure of total mass 2π. Then $H_1(S^1,\mathbb{R})$ is a Hilbert space, with an orthonormal basis given by

$$\{\tfrac{1}{2\pi}\} \cup \{ \frac{\sin n\theta}{\sqrt{\pi(1+n^2)}} , \frac{\cos n\theta}{\sqrt{\pi(1+n^2)}} : n \in N\} .$$

The map $f : \mathbb{R} \to \mathbb{R}$ defined by $f(x) = 2x + \frac{x^2}{2}$ induces a map $H_1(f) : H_1(S^1,\mathbb{R}) \to H_1(S^1,\mathbb{R})$ which is of differentiability class $\mathcal{C}^\infty$. Now, $H_1(f)$ is not a diffeomorphism on $H_1(S^1,\mathbb{R})$. However, if we define $U =\{s \in H_1(S^1,\mathbb{R}) : |s(\theta)| < \frac{3}{2}\ \forall\ \theta \in S^1\}$, then U is open in $H_1(S^1,\mathbb{R})$, and $H_1(f)$ is a diffeomorphism on U.

We will abbreviate $H_1(f)$ with φ.

We now introduce a new metric on U by pulling the usual metric back via φ. Thus, if $v \in H_1(S^1,\mathbb{R})$, and $s \in U$, then

$$\langle v,v\rangle_s = \langle D\varphi_s(v), D\varphi_s(v)\rangle_1 = \langle(2+s)v,(2+s)v\rangle_1 = \int_0^{2\pi}(2v+sv)^2 d\theta + \int_0^{2\pi}(2v'+s'v+sv')^2 d\theta .$$

Let $s_n = \dfrac{\sin n\theta}{\sqrt{\pi(1+n^2)}}$ and $s_0 = 0$. Then each s_n is contained in U, as is s_0, and $\lim_{n\to\infty} s_n = s_0$. The particular v we are interested in is the constant function 1.

We have:

$$\langle 1,1\rangle_{s_n} = \int_0^{2\pi}(2+s_n)^2 d\theta + \int_0^{2\pi}(s_n')^2 d\theta$$

$$= \int_0^{2\pi} 4d\theta + 4\int_0^{2\pi} s_n(\theta)d\theta + \int_0^{2\pi}((s_n)^2+(s_n')^2)d\theta$$

$$= 8\pi + 0 + 1 = 8\pi + 1$$

And: $$\langle 1,1\rangle_{s_0} = \int_0^{2\pi}(2)^2 d\theta = 8\pi$$

Now, if the operator-valued function $(D\varphi_s)^*\circ D\varphi_s$ were weakly continuous from U to $L(H_1(S^1,R),H_1(S^1,R))$ with the topology of uniform weak convergence on bounded sets, then evaluation at the constant function 1 would induce a weakly continuous function from U to $H_1(S^1,R)$ given by $s \longmapsto (D\varphi_s)^*\circ D\varphi_s(1)$. And finally, we could compose this with the continuous linear functional on $H_1(S^1,R)$ to conclude that the real-valued function on U given by $s \longmapsto \langle (D\varphi_s)^*\circ D\varphi_s(1),1\rangle = \langle 1,1\rangle_s$ is continuous. But as we have just seen above, this function is not continuous since $\lim_{n\to\infty}\langle 1,1\rangle_{s_n} \neq \langle 1,1\rangle_{s_0}$. □

The final elementary result about Bw* manifolds which we will mention is a theorem about embedding a connected Bw* manifold in its model space. It is of course a standard and well-known result of Banach manifold theory that any paracompact Hilbert manifold admits a closed embedding into Hilbert space. Kuiper and Terpstra-Kippler have proved a generalization of this theorem which works for manifolds modeled on any Banach space E for which there exists a split linear embedding of $E \oplus E$ into E (the proof is a carefully reworked version of the Hilbert space case —— see [23] for details). An analogous result holds for Bw* manifolds. However, first we need a couple of technical observations:

<u>7.15 Lemma</u>. Let M be a Bw* manifold, X the model space for M, and $x \in M$. Then there exists a chart (U,φ) at x such that φ extends to a homeomorphism of $\overline{U}$ (the closure of U in M) with $\overline{\varphi(U)}$ (the closure of $\varphi(U)$ in X).

Proof. Let (V,φ) be a chart at x. Since M is regular, there is an open neighborhood W of x such that $\overline{W} \subset V$. Since X is regular and $\varphi(W)$ is open in X, there is an open neighborhood U of x such that $\overline{\varphi(U)}$, the closure of $\varphi(U)$ in X, is contained in $\varphi(W)$. Thus the closure of $\varphi(U)$ in X is the same as the closure of $\varphi(U)$ in $\varphi(V)$. And, since $\overline{U} \subseteq \overline{W} \subseteq V$, the closure of U in M is the same as the closure of U in V. Since φ is a homeomorphism of V with $\varphi(V)$, the conclusion is immediate. □

7.16 Remark. If (U,φ) is a chart with the property described in the previous lemma, and W is an open subset of U, then (W,φ) has the same property. Thus, if M is a Bw* manifold, there is a countable locally finite covering $\{(U_n,\varphi_n) \mid n \in N\}$ of M by charts such that φ_n extends to a homeomorphism of $\overline{U_n}$ with $\overline{\varphi(U_n)}$.

7.17 Theorem. Let M be a C^k Bw* manifold modeled on a Bw* space X. Assume that there exists a split linear embedding of $X \oplus X$ into X. Then there exists a closed C^k embedding ρ of M into X.

Proof. For the sake of simplicity we will assume that X is a separable infinite-dimensional weak Hilbert space. The reader who wishes to see the general case can work through Kuiper and Terpstra-Keppler ([23]), making the slight changes necessary to fit the Bw* case.

Let $\{(U_n,\varphi_n) \mid n \in N\}$ be a countable locally finite covering of M by non-empty charts such that φ_n extends to a homeomorphism of $\overline{U_n}$ with $\overline{\varphi(U_n)}$. Let $\{W_n : n \in \mathbf{N}\}$ be a sequence of non-empty open sets in M such that $\overline{W_n} \subset U_n$ and $\bigcup_{n\in N} W_n = M$. For each $n \in N$, let $f_n : M \longrightarrow [0,n]$ be a C^k function such that $f_n^{-1}(n) = \overline{W_n}$ and such that $\operatorname{supp}(f_n) \subset U_n$. Define $Y = \ell_2(\mathbb{R} \oplus X)$. Then Y is a separable infinite-dimensional Hilbert space, so Y is isomorphic to X. We define a map $\rho : M \longrightarrow Y$ as follows:
$\rho(x) = (f_1(x), f_1(x)\varphi_1(x), f_2(x), f_2(x)\cdot\varphi_2(x), \ldots)$. Since, locally,

only a finite number of the factors of ρ are non-zero, ρ is C^k. For each $n \in N$, let $p_n : Y \longrightarrow X$ be projection onto the nth copy of X. Then $p_n \circ \rho|_{\overline{W}_n} = \varphi_n$, which implies that $\rho|_{\overline{W}_n}$ is a homeomorphism onto its image. Furthermore, for each $x \in W_n$, $p_n \circ D\rho(x) = D(p_n \circ \rho)(x) = D\varphi_n(x)$ is an isomorphism, which implies that $D\rho(x)$ is an embedding (since we are dealing with Hilbert space, we don't have to check that the image of $D\varphi(x)$ is a complemented subspace of Y, though that would follow straightforwardly from an examination of $\rho|_{W_n}$).

So to conclude our proof, we need only show that ρ is injective and that $\rho(M)$ is weakly closed in Y. To see that ρ is injective, let y and z be points such that $\rho(y) = \rho(z)$. Choose an integer $q \in N$ such that $y \in W_q$. Since the qth component of $\rho(z)$ is $(f_q(z), f_q(z) \cdot \varphi_q(z))$, and since $\rho(z) = \rho(y)$, we conclude that $f_q(z) = f_q(y) = q$, and hence that $z \in \overline{W}_q$. But since $\rho|_{\overline{W}_q}$ is injective, z must equal y.

To see that $\rho(M)$ is closed in Y, it suffices to show that $\rho(M)$ is sequentially closed. Let $\{x_n\}_{n \in N}$ be a sequence of points in M such that $\{\rho(x_n)\}_{n \in N}$ converges in Y. Since $\{\rho(x_n)\}_{n \in N}$ converges in Y, there is a positive integer $r \in N$ such that $\|\rho(x_n)\| \le r \ \forall n \in N$. Then, for each $m > r$, the mth component of $\rho(x_n)$, $(f_m(x_n), f_m(x_n) \cdot \varphi_m(x_n))$ must be bounded by r in norm, which implies that $x_n \notin W_m$. Thus each x_n is contained in $\bigcup_{m=1}^{r} W_m$. So at least one of the sets $W_1, \ldots, W_r$ must contain an infinite number of terms of the sequence. By choosing a subsequence if necessary, we may assume there is an integer m_0 such that $x_n \in W_{m_0}$ for all $n \in N$. Now, since $\{\rho(x_n)\}_{n \in N}$ is Cauchy in Y, $\{p_{m_0} \circ \rho(x_n)\}_{n \in N}$ is Cauchy in X. But $p_{m_0} \circ \rho(x_n) = \varphi_{m_0}(x_n)$. Let $z_0 = \lim_{n \to \infty} \varphi_{m_0}(x_n)$. Since $z_0 \in \overline{\varphi_{m_0}(W_{m_0})} \subset \varphi_{m_0}(U_{m_0})$, we may define $x_0 = \varphi_{m_0}^{-1}(z_0)$, and we have that $\lim_{n \to \infty} x_n = x_0$. Thus $\lim_{n \to \infty} \rho(x_n) = \rho(x_0)$ is a point in $\rho(M)$, and we have that $\rho(M)$ is closed in Y. □

7.18 Remark. In view of the above theorem it is natural to ask whether every connected infinite-dimensional Bw* manifold can be embedded in its model. The answer to this question is: NO. Elworthy has constructed a simple example of a Banach manifold modeled on James' space which cannot be embedded in its model space, and an examination of this example reveals that it has the structure of a smooth Bw* manifold. Refer to Kuiper and Terpstra-Keppler [23] for details.

We now return to a discussion of Finsler structures (Riemannian metrics):

7.19 Definition. Let U be an open subset of a Bw* space X. A Bw* Finsler structure on $J(U)$ is a Finsler structure such that, for each compact subset $C \subset U$, there exist constants $k_1, k_2 > 0$ such that, for each $x \in C$ and $v \in X$, $k_1\|v\| \leq \|v\|_x \leq k_2\|v\|$.

7.20 Lemma. Let U, W be open subsets of a Bw* space X, and assume $W \Subset U$. If $J(U)$ has a Bw* Finsler structure, then the restriction of this Finsler structure to $J(W)$ is a Bw* Finsler structure on $J(W)$.

7.21 Lemma. Let U, V be open subsets of a Bw* space X, $\varphi : V \longrightarrow U$ a C^1 diffeomorphism. Assume that $J(U)$ has a Bw* Finsler structure. Then φ induces a Bw* Finsler structure by pull-back, where $\|v\|_x$ is defined to be $\|D\varphi_x(v)\|_{\varphi(x)}$ for each $x \in V$, $v \in X$.

7.22 Definition. Let M be a C^k Bw* manifold. A Bw* Finsler structure on $J(M)$ is a Finsler structure on $J(M)$ such that, for each chart (U,φ) on M, the induced Finsler structure on $J(\varphi(U))$ is a Bw* Finsler structure.

Recall that Theorem 7.12 showed the existence of a Bw* Finsler structure on $J(M)$ for each C^k Bw* manifold M with $k \geq 2$.

7.23 Lemma. Let M be a C^k Bw* manifold such that $J(M)$ has a Finsler structure. Assume that, for each $x \in M$, there is a chart

(U_x, φ_x) containing x such that the induced Finsler structure on $J(\varphi_x(U_x))$ is a Bw* Finsler structure. Then the structure on $J(M)$ is a Bw* Finsler structure.

Proof. Let (U, φ) be a chart on M, let A be a compact subset of U, and let X be the model space for M. We must show that there exist constants $c, k > 0$ such that, in the induced Finsler structure on $\varphi(U)$, $c\|v\| \le \|v\|_{\varphi(x)} \le k\|v\|$ for each $x \in A$, $v \in X$.

For each $x \in A$, choose a chart (U_x, φ_x) containing x which induces a Bw* Finsler structure on $\varphi_x(U_x)$, and let W_x be an open neighborhood of x such that $\overline{W}_x \subset U_x$. Since $\{W_x\}_{x \in A}$ is an open covering of A, there is a finite set of points $\{x_i, \ldots, x_m\}$ such that $\{W_{x_i}, \ldots, W_{x_m}\}$ covers A. For simplicity of notation, write W_i for W_{x_i}, φ_i for φ_{x_i}, etc. Let $A_i = A \cap \overline{W}_i$. Then A_i is a compact subset of U_i. Since (U_i, φ_i) induces a Bw* Finsler structure on $\varphi_i(U_i)$, there are positive constants c_i, k_i such that $c_i\|v\| \le \|v\|_{\varphi_i(x)} \le k_i\|v\|$ for all $x \in A_i$, $v \in X$.

Consider the transition function $\varphi_i \circ \varphi^{-1} : \varphi(U_i \cap U) \to \varphi_i(U_i \cap U)$. Since $\varphi_i \circ \varphi^{-1}$ is a diffeomorphism, $D(\varphi_i \circ \varphi^{-1})(\varphi(A_i))$ and $\{\ell^{-1} | \ell \in D(\varphi_i \circ \varphi^{-1})(\varphi(A_i))\}$ are compact. Thus there is a positive integer n_i such that $\frac{1}{n_i}\|v\| \le \|D(\varphi_i \circ \varphi^{-1})_{\varphi(x)}(v)\| \le n_i\|v\|$ for all $x \in A_i$, $v \in X$. So, if $x \in A_i$, $v \in X$, then

$$\|v\|_{\varphi(x)} = \|D(\varphi_i \circ \varphi^{-1})_{\varphi(x)}(v)\|_{\varphi_i(x)} \le k_i\|D(\varphi_i \circ \varphi^{-1})_{\varphi(x)}(v)\| \le k_i n_i\|v\| ,$$

and

$$\|v\|_{\varphi(x)} = \|D(\varphi_i \circ \varphi^{-1})_{\varphi(x)}(v)\|_{\varphi_i(x)} \ge c_i\|D(\varphi_i \circ \varphi^{-1})_{\varphi(x)}(v)\| \ge \frac{c_i}{n_i}\|v\| .$$

So if we define $k = \max\{k_1 n_1, \ldots, k_m n_m\}$ and $c = \min\{\frac{c_1}{n_1}, \ldots, \frac{c_m}{n_m}\}$, we are done. □

A Finsler structure on $J(M)$ induces a metric on $J(M)$, where the distance between two points x and y in $J(M)$ is defined to be the infimum of the lengths of all differentiable paths between x and y if x and y are in the same component of $J(M)$, or ∞ if x and y are in different components of $J(M)$.

7.24 Theorem. Let M be a connected C^k Bw* manifold with a Bw* Finsler structure on $J(M)$, and let A be a compact subset of M. Then A is bounded (in the induced metric) in $J(M)$.

Proof. For each $x \in A$, choose a chart (U_x, φ_x) which contains x and which has the property that $\varphi_x(U_x)$ is a convex subset of the model space X for M. For each $x \in A$, choose an open neighborhood W_x of x such that $\overline{W}_x \subset U_x$. Since the sets $\{W_x\}_{x \in A}$ form an open covering of A, there exists a finite set of points $\{x_1, \dots, x_n\}$ in A such that $\{W_{x_1}, \dots, W_{x_n}\}$ covers A. For simplicity, write W_i for W_{x_i}, φ_i for φ_{x_i}, etc.

For each $1 \le i \le n$, let $A_i = A \cap \overline{W}_i$. Then A_i is a compact subset of U_i, and $A = \bigcup_{i=1}^{n} A_i$. Let C_i be the closed convex hull of $\varphi(A_i)$ in $\varphi(U_i)$. Then C_i is a compact convex subset of $\varphi(U_i)$. So we have the existence of a constant $k_i > 0$ such that $\|v\|_x \le k_i\|v\|$ for each $x \in C_i$ and $v \in X$, where $\| \ \|_x$ is the norm induced on $T_x(\varphi_i(J(U_i)))$ by the Finsler structure on $J(M)$, and $\| \ \|$ is the norm on $E = X''$. Thus the diameter of C_i in the induced Finsler structure is $\le k_i$ multiplied by the diameter of C_i in the normed space E. Since C_i is compact in X, it is bounded in E, and hence it follows that A_i is bounded in $J(M)$. Since A is the union of a finite number of bounded subsets of $J(M)$, we conclude that A is bounded in $J(M)$. □

7.25 Proposition. Let M be a C^k Bw* manifold with a Bw* Finsler structure on $J(M)$, K a nonempty compact subset of M, and C a nonempty closed subset of M such that $K \cap C = \emptyset$. Then, in the induced metric ρ on $J(M)$, $\rho(K,C) > 0$.

Proof. Assume, to the contrary, that $\rho(K,C) = 0$. Then there exist sequences $\{x_n\}_{n\in N}$ in K and $\{w_n\}_{n\in N}$ in C such that $\rho(x_n,w_n) < \frac{1}{n}$ for each $n \in N$. Since K is compact we may assume, by replacing these sequences with subsequences if necessary, that there exists $x_0 \in K$ with $\lim_{n\to\infty} x_n = x_0$. Let X be the model space for the component of M which contains x_0, and let (U,φ) be a chart containing x_0 such that $\varphi(U)$ is convex in X. By ignoring at most a finite number of terms in the sequence, we may also assume that $x_n \in U$ for each $n \in N$.

Let $y_n = \varphi(x_n)$ for each $n \geq 0$. Then, since $\{y_n\}_{n\geq 0}$ is a compact subset of $\varphi(U)$, the closed convex hull D of $\{y_n\}_{n\geq 0}$ is also a compact subset of $\varphi(U)$. Let $d(\cdot,\cdot)$ be the metric on X" induced by the norm: then there exists $\delta > 0$ such that $F = \{y \in X | d(y,D) \leq \delta\} \subset \varphi(U)$. Let $V = \{y \in X | d(y,D) < \delta\}$, and let $G = \{y \in X | d(y,D) = \delta\}$. Note that F is compact in $\varphi(U)$, so that $\varphi^{-1}(F)$ is compact in U, and hence closed in $J(M)$. Since V is open in $\varphi(U)$, $\varphi^{-1}(V)$ is open in $J(M)$, which implies that $\varphi^{-1}(G)$ is closed in $J(M)$.

Consider the Finsler structure induced on $J(\varphi(U))$ by φ^*: there exists $k > 0$ such that $\frac{1}{k}\|v\|_y \leq \|v\| \leq k\|v\|_y$ for each $y \in F$, $v \in X$, where $\| \ \|_y$ is the Finsler metric induced on $T_y(J(\varphi(U)))$ by φ^*. Let $\epsilon = \delta/k$. Then I claim that $\rho(x,w) \geq \epsilon$ for each $x \in \varphi^{-1}(D)$, $w \in M-\varphi^{-1}(V)$. To see this, let $\alpha: I \to J(M)$ be a C^1 curve such that $\alpha(0) = x$ and $\alpha(1) = w$, and let $t_0 = \inf\{t | \alpha(t) \notin \varphi^{-1}(V)\}$. Since $\varphi^{-1}(G)$ is closed in $J(M)$, it follows that $\alpha(t_0) \in \varphi^{-1}(G)$. Thus $\int_0^1 \|\dot{\alpha}(s)\|_{\alpha(s)} ds \geq \int_0^{t_0} \|\dot{\alpha}(s)\|_{\alpha(s)} ds \geq \frac{1}{k}\int_0^1 \|(\varphi\circ\alpha)'(s)\| ds \geq \delta/k = \epsilon$.

Choose $n_0 \in N$ such that $1/n_0 < \epsilon$. Then $w_n \in \varphi^{-1}(V)$ for each $n \geq n_0$. Let $z_n = \varphi(w_n)$ for $n \geq n_0$: then it is immediate that $\rho(x_n,w_n) \geq \frac{1}{k}\|y_n - z_n\|$, which implies that $\lim_{n\to\infty}\|y_n - z_n\| = 0$. Since $\lim_{n\to\infty} y_n = y_0$ in X, it follows that $\lim_{n\to\infty} z_n = x_0$ in X, and hence that $\lim_{n\to\infty} w_n = x_0$ in M, i.e. $x_0 \in K \cap C$. Contradiction. □

We next introduce the concept of a $\mathcal{U}^k$ manifold:

<u>7.26 Definition</u>. A $\mathcal{U}^k$ Bw* manifold M is a Bw* manifold together with a covering of M by charts $\{(U_\lambda,\varphi_\lambda) | \lambda\in\Lambda\}$ such that $\varphi_\beta \circ \varphi_\alpha^{-1}$ is a $\mathcal{U}^k$ map for each $\alpha,\beta \in \Lambda$.

We define $\mathcal{U}^\infty$ maps between $\mathcal{U}^\infty$ manifolds in the natural way. Note that Lemma 7.4 implies the existence of $\mathcal{U}^\infty$ partitions of unity, since any C^∞ real-valued function is automatically $\mathcal{U}^\infty$ by Corollary 6.12. Also, the analogous embedding theorem to 7.17 for $\mathcal{U}^\infty$ manifolds is immediate.

The tangent bundle $T(M)$ to a $\mathcal{U}^\infty$ manifold M is also a $\mathcal{U}^\infty$ manifold (indeed, the definition of $\mathcal{U}^k$ maps for $k > 1$ was made in such a way as to ensure this result). Whether the cotangent bundle $T^*(J(M))$ possesses any structure in addition to that of a Banach manifold is less clear. Recall from Proposition 3.55 however, that the natural model space for $T^*(M)$ is not a D-space, and hence that any differentiable structure on $T^*(M)$ itself would be at best extremely complicated to develop.

An obvious advantage to $\mathcal{U}^\infty$ manifolds and maps over C^∞ manifolds and maps is that we have an inverse function theorem, and hence also an implicit function theorem.

7.27 Implicit Function Theorem (global form). Let M and N be $\mathcal{U}^\infty$ manifolds, $f : M \to N$ a $\mathcal{U}^\infty$ map, and $y_0 \in N$. Assume that, for each $x \in f^{-1}(y_0)$, $Df(x) : T_x(M) \to T_{y_0}(N)$ is surjective, and that $\mathrm{Ker}(Df(x))$ is a complemented subspace of $T_x(M)$. Then $f^{-1}(y_0)$ is a closed $\mathcal{U}^\infty$ submanifold of M.

Proof. This result follows from the local version of the theorem in exactly the same way as the Banach manifold version. Details are left to the reader. □

7.28 Definition. Let M be a C^k Bw* manifold. A Bw* Finsler structure on $J(M)$ will be called complete if bounded subsets of each component of $J(M)$ have compact closure in M.

7.29 Lemma. Let M and N be $\mathcal{U}^\infty$ manifolds, $f : M \to N$ a $\mathcal{U}^\infty$ closed embedding of M into N. Assume that $J(N)$ has a complete Bw* Finsler structure. Then the pull-back of this Finsler structure to $J(M)$ is a complete Bw* Finsler structure on $J(M)$.

Proof. Note that the induced metric on $J(M)$ has the property that, for each $y,z \in J(M)$, $\rho_M(y,z)$ (the distance between y and z in $J(M)$) is $\geq \rho_N(y,z)$. So assume that A is bounded in $J(M)$. Then A is also bounded in $J(N)$, and it follows that the closure of A in N is compact. But since M is a closed subset of N, this means that the closure of A in M is compact.

So we need only show that the Finsler structure on $J(M)$ is a Bw* Finsler structure, and we will do this by showing that the Finsler structure satisfies the condition of Lemma 7.23.

Let $x \in M$. Choose a chart (V,ψ) on N which contains $f(x)$, and such that $\psi(f(x)) = 0$ in the model space Z for N. Choose a chart (W,φ) on M which contains that $f(W) \subset V$ and such that $\varphi(x) = 0$ in the model space X for M. Let $g = \psi \circ f \circ \varphi^{-1} : \varphi(W) \to Z$. Note that g is $\mathcal{U}^\infty$, and $Dg(0)$ is a split linear embedding of X into Z. Choose a linear complement $\tilde{X}$ for $Dg(0)(X)$ in Z, and define $h : \varphi(W) \oplus \tilde{X} \to Z$ by $h(v,w) = g(v) + w$. Then h is $\mathcal{U}^\infty$, and $Dh(0,0)$ is an isomorphism. Thus there is an open neighborhood $\tilde{U}$ of 0 in X and an open neighborhood $\tilde{Y}$ of 0 in $\tilde{X}$ such that $h|_{\tilde{U}\oplus\tilde{Y}}$ is a $\mathcal{U}^\infty$ diffeomorphism. Hence $h^{-1} \circ \psi$ is a chart on N, which implies that the induced Finsler structure on $J(\tilde{U}) \oplus J(\tilde{Y})$ is a Bw* Finsler structure, and hence the restriction of this Finsler structure to $J(\tilde{U})$ is also a Bw* Finsler structure. But if we define U to be $\varphi^{-1}(\tilde{U})$, then this induced Finsler structure on $J(\tilde{U})$ is the Finsler structure which results by pulling back the Finsler structure on $J(M)$ via the chart (U,φ). By Lemma 7.23, we are done. $\square$

Let M be a connected $\mathcal{U}^\infty$ Bw* manifold, X the model space for M, and $E = X''$, and let us return for a moment to the consideration of Theorem 7.17. Even if M cannot be embedded in its model space, it is still possible to embed M in a Bw* space as follows: the space X^N is a locally convex space, and $(B_0(1))^N$ is a compact subset of X^N, where $B_0(1)$ is the closed unit ball in X. Thus we can use $(B_0(1))^N$ and X^N to generate a Bw* space as in Theorem 2.6, and we will denote this space by Z. Obviously, Z

consists of the bounded sequences in X , and it is easy to check that, as a Banach space, $Z'' = \ell_\infty(E)$. Thus it is natural to denote Z by $\ell_\infty(X)$.

An examination of the proof of Theorem 7.17 should make it clear that the proof can be used without modification to obtain a $\mathcal{U}^\infty$ embedding of M into $\ell_\infty(\mathbb{R} \oplus X)$. As a consequence of Lemma 7.29 and the existence of this embedding, we have the following:

<u>7.30 Theorem</u>. Let M be a $\mathcal{U}^\infty$ Bw* manifold. Then J(M) admits a complete Bw* Finsler structure.

<u>Proof</u>. We may assume that M is connected.

Let X be the model space for M , and let $E = X''$. By the above remarks, we know there is a $\mathcal{U}^\infty$ closed embedding of M into $\ell_\infty(\mathbb{R} \oplus X)$. Now, any norm on $\ell_\infty(\mathbb{R} \oplus E)$ induces a "flat" complete Bw* Finsler structure on $\ell_\infty(\mathbb{R} \oplus E)$. But the above lemma implies that the pull-back of this Finsler structure to J(M) is a complete Bw* Finsler structure. □

Let M be a connected $\mathcal{U}^\infty$ manifold, and assume that J(M) has a complete Bw* Finsler structure. Note that a subset $A \subset M$ is bounded in J(M) ⇔ the closure of A in M is compact. This looks very much like the characterization of bounded subsets in a complete finite-dimensional Riemannian manifold. Thus it begins to seem natural to conjecture that there is some kind of infinite-dimensional analogue to the Hopf-Rinow theorem. In fact, if we were dealing with a connected Bw* manifold M , and if we had a Finsler structure on J(M) and a geodesic spray which generated a weakly continuous geodesic flow defined on an open subset of T(M) , then we would be able to prove the following conjecture in essentially the same manner as the Hopf-Rinow theorem is proved:

<u>7.31 Conjecture</u>. The following four statements are equivalent:

(a) J(M) is a complete metric space.

(b) Closed and bounded subsets of J(M) have compact closure in M .

(c) For some $x \in M$, all geodesics from x are infinitely extendable.

(d) All geodesics are infinitely extendable.

And each of the above statements implies:

(e) Any $x,y \in M$ may be joined by a geodesic λ such that $\text{length}(\lambda) = \rho(x,y)$.

The first problem in any attempt to prove this conjecture is that, except for the case of Riemannian metrics on Hilbert manifolds, there is no general technique for obtaining a geodesic spray from a Finsler structure. So let us restrict our attention for a moment to a Bw* Riemannian metric on a weak Hilbert manifold. Then the geodesic spray can be calculated locally in terms of the local representation of the Riemannian metric as a map into the positive definite symmetric operators on Hilbert space. However, if we recall Example 7.14, we see that this local representation of the metric as a map into the positive definite operators is not in general weakly continuous, even though in this example the geodesic spray is $\mathcal{U}^\infty$ and has a $\mathcal{U}^\infty$ flow (that this flow is $\mathcal{U}^\infty$ follows since the metric arises as the pull-back of the flat metric on Hilbert space via a $\mathcal{U}^\infty$ diffeomorphism). So, before attempting to prove the above conjecture, even for the case of Riemannian metrics on weak Hilbert manifolds, it will be necessary to find an abstract condition on the local representation of a Riemannian metric which will ensure that the geodesic spray is $\mathcal{U}^\infty$, subject to the additional requirement that such metrics be constructible on every weak Hilbert manifold.

8. CRITICAL POINT THEORY ON Bw* MANIFOLDS

This chapter is a treatment of the Lusternik-Schnirelman theory of critical points, with the aim of applying this theory to the calculus of variations in the next chapter. We will assume a familiarity with Lusternik-Schnirelman theory on finite-dimensional manifolds and on infinite-dimensional Banach manifolds. The reader who is not familiar with Lusternik-Schnirelman theory on Banach manifolds should refer to Palais [35] for details of the conventional results of critical point theory which we quote here without proof, and also for comparison with the present treatment.

8.1 Definition. Let M be a Banach manifold, $A \subset M$ an arbitrary subset. The subset A will be said to have Lusternik-Schnirelman category m in M (we will write $\mathrm{cat}(A;M) = m$) if A can be covered by m (but no fewer) closed subsets A_i, $1 \le i \le m$, each of which is "contractible to a point in M" (i.e. for each $1 \le i \le m$, the inclusion map $A_i \longrightarrow M$ is homotopic to a map which sends all of A_i to one point). We will define the category of M (written $\mathrm{cat}(M)$) to be $\mathrm{cat}(M;M)$.

The original, and most basic, result of Lusternik-Schnirelman theory is the following:

8.2 Theorem. Let M be a compact C^2 manifold, $f : M \longrightarrow R$ a C^2 function. Then f has at least $\mathrm{cat}(M)$ critical points.

Proof. See [35].

The proof of Theorem 6.2 depends in an essential way upon the compactness of M. When M is assumed to be merely locally compact, the theorem is no longer true in the generality stated above. What we must do in this case is to restrict our attention to functions which have some kind of built-in compactness property to compensate for the global non-compactness of M. One standard assumption of this sort is that f be proper (i.e. $f^{-1}(K)$ is compact for each compact set $K \subset R$). We then have the following:

8.3 Theorem. Let M be a locally compact C^2 manifold, $f : M \longrightarrow R$ a C^2 function which is proper and bounded below. Then f has at least $\operatorname{cat}(M)$ critical points, at least one of which is a minimum for f.

Proof. It will shortly be shown that this is a special case of Theorem 8.5. A separate proof will therefore not be presented. □

When we attempt to generalize the above theorems to functions defined on infinite-dimensional manifolds, a problem becomes apparent: namely, since the manifolds in question cannot be locally compact, functions on these manifolds are never proper.

A solution to this problem was proposed by J. T. Schwartz in [38]. He pointed out that Theorem 8.3 can be generalized to include those functions on infinite-dimensional Hilbert manifolds which satisfy Condition (C) of Palais-Smale.

We recall Condition (C):

8.4 Definition. Let M be a Banach manifold with a Finsler structure, $f : M \longrightarrow R$ a C^1 function. Then f is said to satisfy Condition (C) if, for each sequence $\{x_n\}_{n\in N}$ of points in M such that $\{f(x_n)\}_{n\in N}$ is a bounded subset of R and $\lim_{n\to\infty} \|Df(x_n)\| = 0$, there exists a point $x \in M$ and a subsequence of $\{x_n\}_{n\in N}$ such that the subsequence converges to x.

Note that the continuity of $\|Df\|$ as a real-valued function on M implies that the point x must be a critical point of f.

Schwartz' theorem was subsequently extended by Palais in [32] and [35] to the following:

8.5 Theorem. Let M be a complete C^{2-} Finsler manifold (i.e. a Banach manifold with a complete Finsler structure), $f : M \longrightarrow R$ a C^1 function which is bounded below and satisfies

Condition (C). Then f assumes its minimum and has at least $\mathrm{cat}(M)$ critical points.

Proof. See [35]. □

This will be the starting point for our discussion. We will first examine Condition (C) carefully to see that Theorem 8.5 really is a generalization of Theorem 8.3, and then develop a natural alternative to Condition (C) on Bw* manifolds.

8.6 Lemma (Palais-Smale). Let M be a C^1 Finsler manifold, $f : M \longrightarrow \mathbb{R}$ a C^1 function which satisfies Condition (C), and $K = \{x \mid \|Df(x)\| = 0\}$. Then $f|_K$ is proper.

Proof. It is sufficient to show that $K \cap f^{-1}([a,b])$ is compact for each $a,b \in \mathbb{R}$. But since $\|Df\|$ is identically zero on K, Condition (C) implies that each sequence in $K \cap f^{-1}([a,b])$ has a convergent subsequence. □

8.7 Lemma. Let M be a complete connected C^{2-} Finsler manifold, $f : M \longrightarrow \mathbb{R}$ a C^1 function which is bounded below and which satisfies Condition (C). Then $f^{-1}((-\infty,a])$ is a subset of M which is bounded, in the induced metric on $J(M)$, for each $a \in \mathbb{R}$.

Proof. We may assume that the minimum of f is zero, and in this case it is clearly sufficient to prove the theorem for $a > 0$.

Let K be the set of critical points of f, and let $K_a = K \cap f^{-1}([0,a])$. By Lemma 8.6, K_a is compact, and hence bounded in M. Let $C_a = \{x \in M \mid \rho(x,K_a) \le 1\}$. Then C_a is also a bounded subset of M. Finally, let $M^* = M - C_a$, so that M^* is an open submanifold of M. We construct a vector field on M^* as follows: For each $z \in M^*$, choose $v \in T_z(M^*)$ such that $\|v\| < 1$ and $Df_z(v) < -\frac{\|Df_z\|}{2}$. Let χ_z be a locally Lipschitz vector field on M^* such that $\chi_z(z) = v$. By continuity, there exists a neighborhood U_z of z such that $\|\chi_z(w)\| < 1$ for all $w \in U_z$ and $Df_w(\chi_z(w)) < -\frac{\|Df_w\|}{2}$ for all $w \in U_z$.

By choosing a locally Lipschitz partition of unity on M^* subordinate to the open cover $\{U_z\}_{z\in M^*}$, we may construct a locally Lipschitz vector field χ on M^* such that $\|\chi(w)\| < 1$ for all $w \in M^*$ and $Df_w(\chi(w)) < -\frac{\|Df_w\|}{2}$ for all $w \in M^*$.

Note that, since χ is locally Lipschitz, it generates a maximal flow. Let $x \in f^{-1}([0,a]) \cap M^*$, and consider the maximal integral curve γ_x for χ through x, i.e. $\gamma_x : (b,d) \longrightarrow M^*$, $-\infty \le b < 0 < d < \infty$, with $\gamma_x(0) = x$ and $\gamma_x'(r) = \chi(\gamma_x(r))$ for each $r \in (b,d)$. We will show first that $d < \infty$. To see this, note that Condition (C) implies the existence of a positive number ε such that $\|Df_w\| \ge \varepsilon$ for all $w \in f^{-1}([0,a]) \cap M^*$. Next, note that, since $Df_w(\chi(w)) < 0$ for every $w \in M^*$, $\gamma_x(t) \in f^{-1}([0,a])$ for all $t > 0$ (i.e. f is decreasing along γ_x). Thus, if $0 < t < d$, $a \ge f(x) > f(x) - f(\gamma_x(t)) = -\int_0^t (f\circ\gamma_x)'(s)ds = -\int_0^t Df_{\gamma_x(s)}(\gamma_x'(s))ds > (\varepsilon/2)t$ (Note that this depends crucially on the fact that $f(\gamma_x(t)) > 0$).

Thus $t < 2a/\varepsilon$. Since this holds for all $t \in (0,d)$, we conclude that $d \ge 2a/\varepsilon$.

Now, since $\|\gamma_x'(t)\| < 1$ for all $t \in (0,d)$, and since M is complete, $\lim_{t\to d} \gamma_x(t)$ exists and is an element of C_a. Thus $\rho(x,C_a) \le \int_0^d \|\gamma_x'(t)\|dt < d \le 2a/\varepsilon$. Since C_a is a bounded set, and since $\rho(x,C_a) < 2a/\varepsilon$ for every $x \in f^{-1}([0,a])$, we conclude that $f^{-1}([0,a])$ is bounded in M. □

From now on, we will be interested in discussing critical point theory on Bw* manifolds. As we have seen, we need to assume that the manifolds in question have a Finsler structure if we are going to discuss functions which satisfy Condition (C). Not surprisingly, we will want to assume that our Finsler structures are Bw* Finsler structures of the sort introduced in the previous chapter.

8.8 Notation. A complete Bw* manifold will denote a Bw* manifold with a complete Bw* Finsler structure. In addition, if the Bw* manifold is modeled on Hilbert space, the Finsler structure will be assumed to be the norm function on the tangent bundle which arises from a given Riemannian metric.

For the sake of simplicity, all Bw* manifolds will be assumed to be of differentiability class C^∞.

<u>8.9 Theorem</u>. Let M be a complete Bw* manifold, $f : M \longrightarrow \mathbb{R}$ a strongly C^1 function which is bounded below and which satisfies Condition (C). Then $f^{-1}((-\infty,a))$ has compact closure for each $a \in \mathbb{R}$.

<u>Proof</u>. Obvious from Lemma 8.7.

<u>8.10 Corollary</u>. Let M be a locally compact complete Riemannian manifold, $f : M \longrightarrow \mathbb{R}$ a C^1 function which is bounded below. Then f satisfies Condition (C) $\Leftrightarrow$ f is proper.

<u>Proof</u>. If f is proper, f obviously satisfies Condition (C). Conversely, if f satisfies Condition (C), then $f^{-1}(-\infty,a]$ has weakly compact closure for each $a \in \mathbb{R}$. But the weak and strong topologies are the same on locally compact manifolds, so $f^{-1}(-\infty,a]$ has compact closure. Since f is continuous, $f^{-1}(-\infty,a]$ is also closed, and hence is compact. Thus f is proper. □

<u>8.11 Corollary</u>. Let M be a complete Bw* manifold modeled on an infinite-dimensional Bw* space, $f : M \longrightarrow \mathbb{R}$ a strongly C^1 function which is bounded below and satisfies Condition (C). Then f is not weakly continuous.

<u>Proof</u>. Let $a > \min(f)$: if f were weakly continuous, $f^{-1}(-\infty,a)$ would be a non-empty weakly open set. But Theorem 8.9 states that $f^{-1}(-\infty,a)$ has compact closure in the weak topology. Since M is infinite-dimensional, no non-empty open set in M can have compact closure. Thus $f^{-1}(-\infty,a)$ is not open, and so f is not weakly continuous. □

<u>8.12 Corollary</u>. Let M be a complete Bw* manifold with no locally compact components, $f : M \longrightarrow \mathbb{R}$ a strongly C^1 function which is bounded below and satisfies Condition (C). Let $c = \min(f)$.

Then $f(M) = [c,\infty)$.

Proof. By replacing M with a component on which f assumes its minimum, we may assume that M is connected. Let $a \in \mathbb{R}$ and assume $a > c$. Since $f^{-1}(-\infty,a]$ has compact closure, there exists $x \in M$ with $f(x) > a$. Since M is connected, $[c,a] \subset f(M)$. This being true for each $a > c$, we conclude that $[c,\infty) \subset f(M)$. □

8.13 Definition. Let M be a complete Bw* manifold, $f : M \longrightarrow \mathbb{R}$. The function f will be called pseudo-proper if $f^{-1}([a,b])$ has compact closure in M for each $a,b \in \mathbb{R}$.

Note that a pseudo-proper function is never continuous on M . We will next show that Condition (C) is closely related to two simple conditions —— one a global topological condition, the other a local condition in the weak topology. The global condition is pseudo-properness. The local condition we now introduce:

8.14 Definition. Let M be a complete Bw* manifold, $f : M \longrightarrow \mathbb{R}$ a strongly C^1 function. The function f will be called coercive, if, for each weakly convergent sequence $x_n \longrightarrow x$ in M on which $\lim_{n\to\infty} \|Df(x_n)\| = 0$, we have $x_n \xrightarrow{\text{strongly}} x$.

8.15 Lemma. Let M be a complete Bw* manifold, $f : M \longrightarrow \mathbb{R}$ a strongly C^1 function. Assume that f is pseudo-proper and coercive. Then f satisfies Condition (C) .

Proof. Obvious. □

8.16 Theorem. Let M be a complete Bw* manifold, $f : M \longrightarrow \mathbb{R}$ a strongly C^1 function which is bounded below. Assume in addition that $f(A)$ is a bounded subset of $\mathbb{R}$ for each compact subset A of M . Then f satisfies Condition (C) $\Leftrightarrow$ f is pseudo-proper and coercive.

Proof. We need only prove that, if f satisfies Condition (C), then f is coercive.

Let $\{x_n\}$ be a weakly convergent sequence with limit $x \in M$ on which $\|Df(x_n)\| \longrightarrow 0$. If x_n does not converge strongly to x, then there is a strongly open neighborhood U of x and a subsequence $\{y_n\}_{n\in N}$ such that $y_n \notin U$ for each $n \in N$. But $y_n \xrightarrow{\text{weakly}} x$, which implies that $\{x\} \cup \{y_n\}_{n\in N}$ is weakly compact, and therefore that $\{f(y_n)\}_{n\in N}$ is a bounded set of numbers by our initial assumption. Thus we may apply Condition (C) to conclude that there exists a subsequence $\{y_{n_i}\}_{i\in N}$ and a point $y \in M$ such that $y_{n_i} \xrightarrow{\text{strongly}} y$. But $y_{n_i} \xrightarrow{\text{weakly}} x$, so we must have $y = x$. But since $y_{n_i} \notin U$ for all $i \in N$, $\{y_{n_i}\}$ cannot converge strongly to x. Since this is a contradiction, the neighborhood U and sequence $\{y_n\}_{n\in N}$ as defined above cannot exist, and hence $\{x_n\}_{n\in N}$ must converge strongly to x. □

We now have enough perspective on what Condition (C) means in the case of functions defined on complete Bw* manifolds to introduce a weakened version of Condition (C) and to show that it comes from fairly natural considerations.

<u>8.17 Definition</u>. Let M be a complete Bw* manifold, $f : M \longrightarrow R$ a strongly C^1 function. We will say that f satisfies Condition (WC) if, for each sequence $\{x_n\}_{n\in N}$ in M such that $\{f(x_n)\}_{n\in N}$ is bounded and $\lim_{n\to\infty} \|Df(x_n)\| = 0$, there exists a point $x \in M$ and a subsequence $\{x_{n_i}\}_{i\in N}$ such that:

(1) $x_{n_i} \xrightarrow{\text{weakly}} x$

(2) $f(x_{n_i}) \longrightarrow f(x)$

(3) $Df(x) = 0$

<u>8.18 Notation</u>. K will henceforth be used to denote the set of critical points of a strongly C^1 function. Also, for each $a,b \in R$, we will let $K_a = K \cap f^{-1}(a)$, and $K_{a,b} = K \cap f^{-1}([a,b])$.

<u>8.19 Proposition.</u> Let M be a complete Bw* manifold, $f : M \longrightarrow R$ a strongly C^1 function which satisfies Condition (WC).

Then for each $a,b \in \mathbb{R}$, $K_{a,b}$ is compact and $f|_{K_{a,b}}$ is continuous.

Proof. Let $\{x_n\}_{n\in N}$ be a sequence in $K_{a,b}$. Since $\|Df\|$ is identically zero on K , Condition (WC) implies the existence of a critical point x for f , and a subsequence $\{x_{n_i}\}_{i\in N}$, such that $x_{n_i} \longrightarrow x$ and $f(x_{n_i}) \longrightarrow f(x)$. Since $f(x_{n_i}) \longrightarrow f(x)$, and since $\|Df(x)\| = 0$, we see that $x \in K_{a,b}$. Thus $K_{a,b}$ is sequentially closed in M , and hence closed in M . Thus $K_{a,b}$ is a Lindelöf subspace of M . But the existence of the convergent subsequence $\{x_{n_i}\}_{i\in N}$ implies that $K_{a,b}$ is sequentially compact. And we know that a sequentially compact Lindelöf space is compact.

Finally, since $f(x_{n_i}) \longrightarrow f(x)$, we conclude that f is continuous on convergent sequences in $K_{a,b}$. But $K_{a,b}$ is a closed subspace of the sequentially generated space M . Thus $K_{a,b}$ is sequentially generated, and hence f is continuous on $K_{a,b}$. □

8.20 Corollary. The set $f(K)$ of critical values of f is closed in $\mathbb{R}$.

Proof. Since $K_{a,b}$ is compact and f is continuous on $K_{a,b}$, $f(K_{a,b})$ is compact. But $f(K_{a,b}) = f(K) \cap [a,b]$, which implies that $f(K)$ is closed. □

We will next establish a string of results for functions satisfying Condition (WC) which are virtually identical to those we just established for functions which satisfy Condition (C). Furthermore, the proofs will be only slightly more complicated modifications of the ones we just presented. In those cases where there is no difference at all in the proof, we will content ourselves with a reference to the earlier proof. The first will be an analogue to Theorem 8.9:

8.21 Theorem. Let M be a complete Bw* manifold, $f : M \longrightarrow \mathbb{R}$ a strongly C^1 function which is bounded below and satisfies Condition (WC) . Then f is pseudo-proper.

Proof. Let $a \in \mathbb{R}$, with $a > \min(f)$. We must show that $f^{-1}(-\infty,a]$ is bounded in $J(M)$.

We proceed as follows: for each $n \in N$, let $C_n = f^{-1}(-\infty,a] \cap \{x \mid \|Df(x)\| \leq 1/n\}$. Then C_n is strongly closed, and $C_{n+1} \subset C_n$ for each $n \in N$. We assert that there exists $n_0 \in N$ such that C_{n_0} intersects only a finite number of components of M. For, if not, it would be possible to choose a sequence $\{x_n\}_{n \in N}$ such that $x_n \in C_n$ but such that no two points were in the same component of M. Such a sequence could not have a convergent subsequence; but the sequence would have to have a convergent subsequence by Condition (WC). Since this would be a contradiction, we see that there exists $n_0 \in N$ such that C_{n_0} intersects only a finite number of components of M.

Next, I claim that there exists a positive integer $n_1 \geq n_0$ such that C_{n_1} has compact closure in M. To see this, let $M_1,\ldots,M_j$ be the components of M which intersect C_{n_0}. For each $i \in \{1,\ldots,j\}$, I assert there is a positive integer m_i such that $C_{m_i} \cap M_i$ has compact closure. For if not, let $z \in M_i$: it would then be possible to choose a sequence $\{y_n\}_{n \in \mathbf{N}}$ such that $y_n \in C_n \cap M_i$ and such that $\rho(y_n,z) > n$ (recall that, in a connected component, non-compactness is equivalent to unboundedness). Such a sequence could have no convergent subsequence, which would violate Condition (WC). Thus there must exist a positive integer for which C_{m_i} has compact closure. To finish the claim, let $n_1 = \max\{m_1,\ldots,m_j\}$.

The rest of the proof is essentially identical to the proofs of Lemma 8.7 and Theorem 8.9, with C_{n_1} substituted for C_a in the proof of 8.7 (Note that $M^* = M - C_{n_1}$ is a strongly open submanifold of M, so that we can construct a vector field as we did in the proof of 8.7). Therefore, the remainder of the proof will be omitted. □

8.22 Definition. Let M be a complete Bw* manifold, $f : M \longrightarrow \mathbb{R}$ a strongly C^1 function. We will call f weakly coercive if, for each weakly convergent sequence $x_n \longrightarrow x$ in M on which $\lim_{n\to\infty} \|Df(x_n)\| = 0$, we have $f(x_n) \longrightarrow f(x)$ and $\|Df(x)\| = 0$.

Note that the strong continuity of f and Df imply that a coercive function is automatically weakly coercive.

The next two results are natural generalizations of Lemma 8.15 and Theorem 8.16 respectively, with virtually identical proofs:

<u>8.23 Lemma</u>. Let M be a complete Bw* manifold, $f : M \longrightarrow \mathbb{R}$ a strongly C^1 function which is pseudo-proper and weakly coercive. Then f satisfies Condition (WC).

<u>Proof</u>. See 8.15. □

<u>8.24 Theorem</u>. Let M be a complete Bw* manifold, $f : M \longrightarrow \mathbb{R}$ a strongly C^1 function which is bounded below. Assume in addition that $f(A)$ is a bounded subset of $\mathbb{R}$ for each compact subset A of M. Then f satisfies Condition (WC) $\Leftrightarrow$ f is pseudo-proper and weakly coercive.

<u>Proof</u>. See 8.16. □

<u>8.25 Proposition</u>. Let M be a complete Bw* manifold, $f : M \longrightarrow \mathbb{R}$ a strongly C^1 function which is bounded below, satisfies Condition (WC), and is bounded on each compact subset of M. Then K is a closed locally compact subspace of M, and $f|_K$ is both continuous and proper.

<u>Proof</u>. Let x be an accumulation point of K. Since the closure of K is the same as the sequential closure, there exists a sequence $\{x_n\}_{n\in N}$ of points in K such that $\lim_{n\to\infty} x_n = x$. Since f is bounded on compact subsets of M, f is bounded on the sequence. Thus by the weak coercivity of f, $x \in K$. Thus K is sequentially closed, and hence closed, in M.

To see that $f|_K$ is continuous, note that the topology on K is sequentially generated (since it is a closed subspace of a sequentially generated space). Thus it suffices to check the continuity of f on convergent sequences, and the continuity of f on convergent sequences was shown in Proposition 8.19.

Let $[a,b]$ be any finite closed interval. Since $K_{a,b}$ is compact, $f|_K$ is proper. And finally, since $f|_K$ is continuous, $(f|_K)^{-1}((a,b)) = U_{a,b}$ is open in K. But $\overline{U}_{a,b} \subset (f|_K)^{-1}([a,b]) = K_{a,b}$, which implies that K is locally compact. □

A connection between Conditions (C) and (WC), or equivalently between coercivity and weak coercivity, is given by a property which is possessed by nonlinear functionals which arise in certain quasi-linear elliptic variational problems, and which we will therefore call "ellipticity":

8.26 <u>Definition</u>. Let M be a complete Bw* manifold, $f : M \longrightarrow \mathbb{R}$ a strongly C^1 function. The function f will be said to be <u>elliptic</u> if, for each weakly convergent sequence $x_n \longrightarrow x$ in M for which $f(x_n) \longrightarrow f(x)$, we have $x_n \xrightarrow{\text{strongly}} x$.

8.27 <u>Remark</u>. Let M be a complete Bw* manifold $f : M \longrightarrow \mathbb{R}$ an elliptic strongly C^1 function. Then:

(1) f satisfies Condition (C) $\Leftrightarrow$ f satisfies Condition (WC).

(2) f is coercive $\Leftrightarrow$ f is weakly coercive.

Our next goal will be to prove a new version of Theorem 8.5, in which Condition (C) is replaced by Condition (WC). The method of proof will be based upon the standard proof of 8.5, but because it is slightly more complicated we will bother with the details. Our method of proof will employ the usual tool of a pseudo-gradient vector field, so we will quickly review the relevant properties of pseudo-gradient vector fields on Banach manifolds. The interested reader can refer to [35] —— from which our treatment is borrowed —— for details and complete proofs (including a proof of 8.5).

8.28 <u>Definition</u>. Let M be a C^{2-} Finsler manifold, $f : M \longrightarrow \mathbb{R}$ a C^1 function, K the set of critical points of f, and $M^* = M - K$. A vector $v \in T(M^*)_p$ will be called a pseudo-gradient vector for f at p if $\|v\| < 2\|Df(p)\|$ and $Df(p)(v) > \|Df(p)\|^2$. A locally Lipschitz vector field χ on M^* will be called a pseudo-gradient vector field for f if, for each $p \in M^*$, $\chi(p)$ is a pseudo-gradient vector for f at p.

<u>8.29 Remark</u>. Every C^1 function has a pseudo-gradient vector field.

<u>8.30 Notation</u>. Let χ be a pseudo-gradient vector field for f. We will let Φ denote the maximal flow generated by $-\chi$.

For each $p \in M^*$, the map $t \longrightarrow \Phi_t(p)$ is a C^1 map of an open interval $(\alpha(p), \omega(p))$ which contains zero into M^* such that $\Phi_0(p) = p$ and $\Phi_t'(p) = -\chi(\Phi_t(p))$. The maximality of Φ_t is equivalent to the two properties that, for each $p \in M^*$, either $\alpha(p) = -\infty$ or else $\Phi_t(p)$ has no limit point in M^* as $t \longrightarrow \alpha(p)$, and similarly that either $\omega(p) = \infty$ or else $\Phi_t(p)$ has no limit point in M^* as $t \longrightarrow \omega(p)$.

Let $\rho : M \times M \longrightarrow R$ denote the metric induced by the Finsler structure on M, and for $\alpha(p) < t_1 < t_2 < \omega(p)$, let $\ell_p(t_1,t_2)$ denote the length of the C^1 curve $t \longrightarrow \Phi_t(p)$ between t_1 and t_2. Then we have the inequalities of the following theorem:

<u>8.31 Theorem</u>. Let $p \in M^*$, and $\alpha(p) < t_1 < t_2 < \omega(p)$. Then:

(1) $$\int_{t_1}^{t_2} \|Df(\Phi_t(p))\|^2 dt < f(\Phi_{t_1}(p)) - f(\Phi_{t_2}(p))$$

(2) $$\rho(\Phi_{t_1}(p), \Phi_{t_2}(p)) \le \ell_p(t_1,t_2) < 2(t_2-t_1)^{1/2}(f(\Phi_{t_1}(p)) - f(\Phi_{t_2}(p)))^{1/2}.$$

<u>Proof</u>. To see (1), note that $f(\Phi_{t_1}(p)) - f(\Phi_{t_2}(p)) = \int_{t_1}^{t_2} (Df(\Phi_t(p))(\Phi_t'(p))dt > \int_{t_1}^{t_2} \|Df(\Phi_t(p))\|^2 dt$. To see (2), note first that

$$\rho(\Phi_{t_1}(p), \Phi_{t_2}(p)) \le \ell_p(t_1,t_2) = \int_{t_1}^{t_2} \|\Phi_t'(p)\| dt =$$

$$= \int_{t_1}^{t_2} \|\chi(\Phi_t(p))\| dt < 2 \int_{t_1}^{t_2} \|Df(\Phi_t(p))\| dt .$$

Now, from Schwartz' inequality,

$$\int_{t_1}^{t_2} \|Df(\Phi_t(p))\| dt \le (t_2 - t_1)^{1/2} \left(\int_{t_1}^{t_2} \|Df(\Phi_t(p))\|^2 dt\right)^{1/2} ,$$

and so we may apply the first inequality to this equation to conclude that $\ell_p(t_1, t_2) \le (t_2 - t_1)^{1/2} (f(\Phi_{t_1}(p)) - f(\Phi_{t_2}(p)))^{1/2}$. □

<u>8.32 Corollary</u>. Let M be complete. Assume in addition that $\omega(p) < \infty$, and $\{f(\Phi_t(p)) \mid t \in (\alpha(p)(\omega(p))\}$ is bounded below (which would be the case, for instance, if f is bounded below). Then $\lim_{t \to \omega(p)} \Phi_t(p)$ exists in M, and is a critical point of f.

Now we will return to the examination of critical point theory on Bw* manifolds. In what follows, M will denote a complete Bw* manifold, f a strongly C^1 function on M which is bounded below and which satisfies Condition (WC), and χ a pseudo-gradient for f. We will use B to denote $\inf\{f(x) \mid x \in M\}$. The next several results are slight modifications of results from the theory of critical points on Banach manifolds which may be found in [35].

<u>8.33 Lemma</u>. Let $p \in M^*$. Then, as $t \longrightarrow \omega(p)$, $f(\Phi_t(p))$ converges monotonically to a limit $a \in \mathbb{R}$ with $a \ge B$. If $\omega(p) < \infty$ then $\Phi_t(p)$ converges to an element of K_a, while if $\omega(p) = \infty$ then $\Phi_t(p)$ has at least one weak limit point q as $t \longrightarrow \infty$. In the latter case, any such weak limit belongs to K_a, which implies that, in either case, a is a critical value of f. As $t \longrightarrow \alpha(p)$, either $f(\Phi_t(p))$ diverges monotonically to ∞ or else it converges monotonically to a finite limit b. In the latter case, we have a situation similar to the case where $t \longrightarrow \omega(p)$: namely, if $\alpha(p) > -\infty$, then $\Phi_t(p)$ converges to a point of K_b as $t \longrightarrow \alpha(p)$, while if $\alpha(p) = -\infty$, then $\Phi_t(p)$ has at least one weak limit point q as $t \longrightarrow -\infty$. Furthermore, any such weak limit is an element of K_b, which implies that, again, b is a critical value of f.

Proof. We will treat the case $t \longrightarrow \omega(p)$, the case $t \longrightarrow \alpha(p)$ being nearly identical. Because of Corollary 8.32, we need only consider the case where $\omega(p) = \infty$. Let $t > 0$. By (1) of Theorem 8.32, $\int_0^s \|Df(\Phi_t(p))\|^2 dt < f(\Phi_0(p)) - f(\Phi_s(p)) \le f(p) - B$. Taking the limit as $t \longrightarrow \infty$, we have $\int_0^\infty \|Df(\Phi_t(p))\|^2 dt < \infty$, which implies that $\|Df(\Phi_t(p))\|$ cannot be bounded away from zero as $t \longrightarrow \infty$. By Condition (WC), $\Phi_t(p)$ has a weak limit as $t \longrightarrow \infty$. And again by Condition (WC), any such weak limit point must be in K_a. □

8.34 Lemma. Let g be a locally Lipschitz function on $J(M)$ with $0 \le g \le 1$. Assume that there exists $c \in R$ such that $g(x) = 0$ for all $x \in M$ with $f(x) \ge c$, and assume also that g vanishes on a strong neighborhood of K. Define a locally Lipschitz vector field Y on $J(M)$ by $Y(x) = -g(x)\chi(x)$ if $x \in M^*$, $Y(x) = 0$ if $x \in K$. Then the maximal flow ψ_t generated by Y is a one-parameter group of Lipschitz homeomorphisms of $J(M)$, i.e. for each $p \in M$, $\psi_t(p)$ is defined for $t \in (-\infty,\infty)$.

Proof. Let $p \in M$. If $Y(p) = 0$, then $\psi_t(p) = p$ is obviously defined for all $t \in (-\infty,\infty)$, so assume that $Y(p) \ne 0$. This implies that $p \in M^*$ and $f(p) < c$. Let (γ,δ) be the interval on which the curve $\psi_t(p)$ is defined. Then we have a relation between $\psi_t(p)$ and the integral curve $\Phi_t(p)$ to the vector field $-\chi$: namely, there is a C^1 function λ defined on (γ,δ) with range contained in $(\alpha(p),\omega(p))$ such that $\lambda(0) = 0, \lambda'(t) \in (0,1]$ for all $t \in (\gamma,\delta)$, and such that $\psi_t(p) = \Phi_{\lambda(t)}(p)$.

Assume for the moment that $\delta < \infty$, and let $\tilde{\omega}(p) = \lim_{t\to\delta} \lambda(t)$. Then, since $|\lambda'(t)| \le 1$ for all t, $\tilde{\omega}(p) \le \delta < \infty$. By (2) of Theorem 8.31, we see that $\lim_{t\to\omega(p)} \Phi_t(p)$ exists. But this implies that $\lim_{t\to\delta} \Phi_{\lambda(t)}(p)$ exists, and since $\Phi_{\lambda(t)}(p) = \psi_t(p)$, we conclude that $\lim_{t\to\delta} \psi_t(p)$ exists. But this is impossible, since the existence of this limit would imply that we could extend the domain of definition of $\psi_t(p)$ past δ, which would contradict the assumed maximality of (γ,δ) . Thus δ must be ∞ .

To see that $\gamma = -\infty$, assume to the contrary for the moment that $\gamma > -\infty$. Note that, since $Y(p) \neq 0$, $Y(\psi_t(p)) \neq 0$ for all $t \in (\gamma,\delta)$. Thus $f(\psi_t(p)) < c$ for all $t \in (\gamma,\delta)$. Let $\tilde{\alpha}(p) = \lim_{t\to\gamma} \lambda(t)$. Then we may proceed as in the case of $\tilde{\omega}(p)$ to show that $\lim_{t\to\gamma} \psi_t(p)$ exists, and thus to obtain another contradiction. □

8.35 Notation. If $f : M \longrightarrow \mathbb{R}$, we will use f^c to denote $f^{-1}((-\infty,c])$.

8.36 Deformation Theorem. Let $c \in \mathbb{R}$, and let U be any open neighborhood of K_c in M . Then there exists a one-parameter group ψ_t of locally Lipschitz homeomorphisms on $J(M)$ and a positive number ε such that $\psi_1(f^{c+\varepsilon} - U) \subset f^{c-\varepsilon}$.

Proof. Let χ be a pseudo-gradient vector field for f , Φ the flow generated by $-\chi$. Choose an open subset V in M such that $K_c \subset V \subset \overline{V} \subset U$ (if K_c is empty, let $V = \emptyset$) . By Condition (WC), we know there exists $r \in (0,1)$ such that, for each $x \in f^{-1}([c-r,c+r])$ with $x \notin V$, $\|Df(x)\| \geq r$. Let g be a locally Lipschitz function on $J(M)$ such that: $0 \leq g \leq 1$, g vanishes in a strong neighborhood of K_c, $g(x) = 0$ if $f(x) > c+1$, and $g(x) = 1$ for all $x \in f^{c+r}$ such that $x \in M - V$. Let Ψ_t denote the one-parameter group of homeomorphisms of $J(M)$ generated by $-g\chi$.

Consider the two sets $f^{c+r} \cap V$ and $M - U$. If $f^{c+r} \cap V$ is non-empty, let $\delta = \rho(f^{c+r} \cap V, M - U)$, i.e. the distance between the two sets. By Proposition 7.25, we know that $\delta > 0$. If $f^{c+r} \cap V$ is empty, let $\delta = 2r$. Set $\varepsilon = \min\{\delta^2/8, r^2/2\}$.

Let $x \in f^{c+\varepsilon} - U$. I claim that $\Psi_1(x) \in f^{c-\varepsilon}$. To see this, assume to the contrary that $f(\psi_1(x)) > c-\varepsilon$. We have two possible cases to consider: either $\Psi_t(x) \in f^{c+\varepsilon} - V$ for all $t \in [0,1)$, or there exists $t_1 \in (0,1)$ with $\Psi_{t_1}(x) \in V$.

In the first case, since $\Psi_t(x) \in f^{c+\varepsilon} - V$ for all $t \in [0,1]$, we have $\Psi_t(x) = \Phi_t(x)$ for all $t \in [0,1]$. Thus
$f(x) - f(\Psi_1(x)) = -\int_0^1 Df(\Psi_t(x))(\Psi_t'(x))dt = -\int_0^1 Df(\Phi_t(x))(\Phi_t'(x))dt$
$\geq \int_0^1 r^2\, dt \geq 2\varepsilon$. But this implies that $f(\Psi_1(x)) \leq f(x) - 2\varepsilon \leq c-\varepsilon$, which contradicts the assumption that $f(\Psi_1(x)) > c-\varepsilon$, and hence is impossible.

Now let us consider the second case: let t_0 be the smallest $t \in (0,1]$ such that $\Psi_t(x) \in \overline{V}$. Then $\Psi_t(x) = \Phi_t(x)$ for all $t \in [0,t_0]$. Thus, by (2) of Theorem 8.31,
$f(x) - f(\Psi_{t_0}(x)) \geq \delta^2/4t_0 \geq \delta^2/4 \geq 2\varepsilon$, which implies
$f(\Psi_1(x)) \leq f(\Psi_{t_0}(x)) \leq f(x) - 2\varepsilon \leq c-\varepsilon$: another contradiction. Thus this case is also an impossibility, and so we conclude that
$\Psi_1(f^{c+\varepsilon} - U) \subset f^{c-\varepsilon}$. □

<u>8.37 Theorem</u>. For each integer k with $1 \leq k \leq \text{cat}(J(M))$, define $c_k = \text{Inf}\{c \in \mathbb{R} \mid \exists\, A \subseteq f^c \text{ with } \text{cat}(A; J(M)) \geq k\}$
$= \text{Inf}\{c \in \mathbb{R} \mid \text{cat}(f^c; J(M)) \geq k\}$.
Then either $c_k = \infty$ or else c_k is a critical value of f. Moreover, if $c_k = c_{k+1} = \cdots = c_{k+j} < \infty$, then f has at least $j + 1$ critical points on the level c_k.

<u>Proof</u>. If c_k were not a critical value of f, then by Theorem 8.36 there would exist $\varepsilon > 0$ and a one-parameter group Ψ_t of homeomorphisms on M such that $\Psi_1(f^{c+\varepsilon}) \subset f^{c-\varepsilon}$, which would imply that $\text{cat}(f^{c+\varepsilon}; J(M)) = \text{cat}(f^{c-\varepsilon}; J(M)) \leq k - 1$, which would be a contradiction. Thus c_k must be a critical value of f.

To see the second statement, assume that $c = c_k = c_{k+1} = \ldots = c_{k+j}$. If there are an infinite number of critical points of f at level c we are done; so assume that, to the contrary, there are only a finite number, say $x_1,\ldots,x_m$. Then the set of critical points of f on level c is discrete, and since M is a manifold we can choose weakly open neighborhoods U_i of x_i such that $U_i \cap U_j = \emptyset$ if $i \neq j$ and such that each U_i is contractible. Let $U = \bigcup_{i=1}^m U_i$: by Theorem 8.36, there exists $\varepsilon > 0$ and a

one-parameter group Φ_t of homeomorphisms on $J(M)$ such that $\Phi_t(f^{c+\varepsilon} - U) \subset f^{c-\varepsilon}$, which implies that $\mathrm{cat}(f^{c+\varepsilon} - U; J(M)) \leq k-1$. Now, $\mathrm{cat}(U; J(M)) \leq m$, and so we see that
$k + j \leq \mathrm{cat}(f^{c+\varepsilon}; J(M)) \leq \mathrm{cat}(f^{c+\varepsilon} - U; J(M)) + \mathrm{cat}(U; J(M))$
$\leq k - 1 + m$, which in turn yields $m \geq j + 1$. □

8.38 Corollary. f has at least $\mathrm{cat}(J(M))$ critical points.

Proof. If f has an infinite number of critical points we are done, so assume that, to the contrary, f has only a finite number of critical points. It suffices to show that $c_k < \infty$ for each $1 \leq k \leq \mathrm{cat}(J(M))$, for then Theorem 8.37 will imply the existence of at least one critical point for each k. To see this, choose $r \in \mathbb{R}$ such that $r - 1$ is greater than all the critical levels of f. Let g be a locally Lipschitz function on $J(M)$ such that $0 \leq g(x) \leq 1$ for all $x \in M$, $g(x) = 0$ for all $x \in f^{r-1}$, and $g(x) = 1$ for all x with $f(x) \geq r$. Let Ψ_t be the maximal flow generated by $-g\cdot\chi$. Then the first half of the proof of Lemma 8.34 shows that, for each x, $\Psi_t(x)$ is defined for all $t \geq 0$. Let $x_0 \in M$ with $f(x_0) > r$: since f has no critical values in $[r, f(x_0)]$, Condition (WC) implies the existence of $\delta > 0$ such that $\|Df(x)\| \geq \delta$ for all x with $f(x) \in [r, f(x_0)]$. Thus $\frac{d(f \circ \Psi_t(x_0))}{dt} \leq -\delta^2$ for all t such that $f \circ \Psi_t(x_0) \geq r$. This implies that $f \circ \Psi_t(x_0) \leq r$ for all $t \geq \frac{f(x_0)-r}{\delta^2}$. For each $x \in M$, define $\lambda(x)$ by $\lambda(x) = \inf\{t \in \mathbb{R} \mid f \circ \Psi_t(x) \leq r\}$. It is easy to see that λ is a continuous function on $J(M)$. Define a deformation h_t of the identity map on $J(M)$ by $h_t(x) = \Psi_{t\cdot\lambda(x)}(x)$. Then $h_t : I \times J(M) \to J(M)$ is continuous, and $h_1(J(M)) \subset f^r$. Thus $\mathrm{cat}(J(M)) = \mathrm{cat}(f^r; J(M))$, which implies that $c_k \leq r$ for all $1 \leq k \leq \mathrm{cat}(J(M))$. □

8.39 Corollary. f assumes a minimum on M, and assumes a minimum on each component of M.

Proof. Let $m = \inf\{f(x) \mid x \in M\}$. Then $\mathrm{cat}(f^c; J(M)) = 0$ if $c < m$ and $\mathrm{cat}(f^c; J(M)) > 0$ if $c > m$. By theorem 8.37, it follows that m is a critical level of f, and hence that $f^{-1}(m)$ is not empty. To see the other assertion, note that each component of M is itself a complete Bw* manifold on which f is bounded below and satisfies Condition (WC) . □

We next want to find some interesting examples of functions which are bounded below and which satisfy Condition (WC) . By Theorem 8.21, any such function must be pseudo-proper; so, if M is modeled on an infinite-dimensional Bw* space, f cannot be weakly continuous.

<u>8.40 Theorem</u>. Let $F : M \to X$ be a $\mathcal{U}^1$ map from a $\mathcal{U}$-manifold M to a Hilbert space X which is proper. Define $f : M \to R$ by $f(x) = \langle F(x), F(x)\rangle = \|F(x)\|^2$. Then f is bounded below and satisfies Condition (WC) .

<u>8.41 Corollary</u>. Let $i : M \to X$ be an embedding. Then the function $f : M \to R$ defined by $f(x) = \|i(x)\|^2$ is bounded below and satisfies Condition (WC) .

We will not prove these results, since we intend to extend these statements to general Bw* spaces and to prove the generalizations. However, the above statements suggest a way around the difficulty that pseudo-proper functions cannot be continuous: namely, we look for functions on Bw* manifolds which can be represented as the composition of a weakly proper $\mathcal{U}$-map from the manifold to a linear space, followed by a power of the norm function on the linear space. Since the norm function (and hence any power of the norm function) is pseudo-proper, such a composition is automatically pseudo-proper. Since the norm is the simplest pseudo-proper function on a linear space, we thus arrive at a class of nonlinear discontinuous functionals on Bw* manifolds in which the complicated aspects of their nonlinearities are confined to the $\mathcal{U}$-map part of the composition and in which all complications arising from the lack of weak continuity are confined to the norm part.

Since we are interested in functionals which satisfy Condition (WC) , it should be apparent that we will need to restrict our attention to norms on Bw* spaces which satisfy certain properties relative to the weak topology. An analysis of the proof of Theorem 8.40 suggested Property (A) of Definition 8.45 as the key property.

For purposes of pedagogical completeness, a summary of relevant material concerning norms on Banach spaces is next presented. This material is more or less standard, albeit a bit specialized; I am grateful to R. R. Phelps and J. H. M. Whitfield for making me aware of it. The reader is advised to simply think of the important examples of Hilbert spaces and the L^p spaces, and not worry about proofs.

Let E be a Banach space (with a given norm):

<u>8.42 Definition</u>. A duality map for E is a map $f : E \to E'$ (not necessarily continuous) such that $\|x\| = \|f(x)\|$ and such that $\langle f_x, x\rangle = \|x\|^2$ for all $x \in E$, where f_x will be used to denote $f(x)$.

<u>8.43 Definition</u>. The norm on E will be called C^1 if the function $\| \quad \| : E \to \mathbb{R}$ given by $x \mapsto \|x\|$ is C^1 on $E - \{0\}$.

<u>8.44 Remarks</u>. (a) Every Banach space admits a duality map by the Hahn-Banach theorem.

(b) If E admits a duality map which is continuous, then there is only one duality map, and the norm function on E is C^1.

(c) If E has a C^1 norm, then its derivative is given by $D(\| \quad \|)_x = \dfrac{f_x}{\|x\|}$ for all $x \neq 0$. Note that the continuity of $D(\| \quad \|)$ away from zero implies the continuity of f on all of E (including zero), and hence that part (b) implies the uniqueness of the functional f_x for each $x \in E$.

(d) If we let $p > 1$, and we consider the function $\| \quad \|^p$ on E, then $\| \quad \|^p$ is a C^1 function on all of E (including 0), $D(\| \quad \|^p)_x = p\|x\|^{p-1} f_x$ for all $x \neq 0$, and $D(\| \quad \|^p)_0 = 0$.

8.45 Definition. The Banach space E has Property (A) if E admits a duality map f such that: for each sequence $\{x_n\}_{n\in N}$ with $\|x_n\| = 1$ for all $n \in N$ and $\lim_{n,m\to\infty} \langle f_{x_n} - f_{x_m}, x_n - x_m\rangle = 0$, we have $\lim_{n,m\to\infty} \|x_n - x_m\| = 0$.

8.46 Theorem. Assume that E has a C^1 norm. Then the norm on E has Property (A) $\Leftrightarrow$ E is reflexive and the induced dual norm on E' is also C^1.

8.48 Corollary. Let (M,μ) be a measure space, and $1 < p < \infty$. Then the space $L^p(M,R)$ has a C^1 norm and satisfies Property (A).

Proof. For $1 < p < \infty$, $L^p(M,R)$ has a C^1 norm. But the dual space of $L^p(M,R)$ is $L^{p/p-1}(M,R)$. Since $1 < \frac{p}{p-1} < \infty$, $L^{p/p-1}(M,R)$ also has a C^1 norm.

8.49 Theorem. Any reflexive Banach space admits an equivalent norm which is C^1 and which satisfies Property (A).

Proof. See [44]. □

8.50 Definition. A Banach space will be called variational if its norm is C^1 and satisfies Property (A).

Note that, if E is a variational space, then it is automatically reflexive by Theorem 8.46. Thus E can be retopologized into a Bw* space, and in only one way. So, if M is a $\mathcal{U}$-manifold, we may speak of a $\mathcal{U}$-map from M to E, by which we of course mean a map between two Bw* manifolds, namely: M, and E with its uniquely determined Bw* topology.

8.51 Theorem. Let E be a variational Banach space, f the duality map on E, and $\{x_n\}_{n\in N}$ a bounded sequence in E such that $\langle f_{x_n} - f_{x_m}, x_n - x_m\rangle \to 0$ as $n,m \to \infty$. Then $\|x_n - x_m\| \to 0$ as $n,m \to \infty$.

Proof. If $\|x_n\| \to 0$ as $n \to \infty$, we are done. So assume that this is not the case. Then there exists $r > 0$ and a subsequence $\{y_n\}_{n \in N}$ of the original sequence such that $\lim_{n\to\infty} \|y_n\| = r$. In addition, we may assume that none of the terms of the subsequence is zero. Let $z_n = \frac{y_n}{\|y_n\|}$. Since the norm is C^1, part (b) of 8.44 implies that $f_{z_n} = \frac{1}{\|y_n\|} f_{y_n}$ for each $n \in N$. Now,

$$<f_{z_n} - f_{z_m}, z_n - z_m> = < \frac{1}{\|y_n\|} f_{y_n} - \frac{1}{\|y_m\|} f_{y_m}, \frac{y_n}{\|y_n\|} - \frac{y_m}{\|y_m\|} >$$

$$= 2 - \frac{1}{\|y_n\| \cdot \|y_m\|} \{<f_{y_n}, y_m> + <f_{y_m}, y_n>\}$$

$$= 2 + \frac{1}{\|y_n\| \cdot \|y_m\|} <f_{y_n} - f_{y_m}, y_n - y_m> - \frac{\|y_n\|}{\|y_m\|} - \frac{\|y_m\|}{\|y_n\|} .$$

Since $\lim_{n,m\to\infty} <f_{y_n} - f_{y_m}, y_n - y_m> = 0$, and since $\lim_{n\to\infty} \|y_n\| = r$, this implies that $\lim_{n,m\to\infty} <f_{z_n} - f_{z_m}, z_n - z_m> = 0$, and hence that there exists $z_o \in E$ with $\lim_{n\to\infty} z_n = z_o$. By continuity of scalar multiplication, $\lim_{n\to\infty} y_n = \lim_{n\to\infty}(\|y_n\| \cdot z_n) = r \cdot z_o$.

Thus, if $\lim_{n\to\infty} \|x_n\|$ exists, we are done. To see that this limit does in fact exist, let us assume for a moment that it does not exist. Then there would exist two distinct real numbers a, b with $0 \le a < b$ and subsequences $\{y_n\}_{n \in N}$ and $\{z_m\}_{m \in N}$ of our original sequence such that $\lim_{n\to\infty} \|y_n\| = a$ and $\lim_{m\to\infty} \|z_m\| = b$. Then

$$<f_{y_n} - f_{z_m}, y_n - z_m> = <f_{y_n}, y_n> + <f_{z_m}, z_m> - <f_{y_n}, z_m> - <f_{z_n}, y_m> .$$

Let $\varepsilon = \frac{(b-a)^2}{8}$: then, for sufficiently large r and m, $<f_{y_n}, y_n> \ge a^2 - \varepsilon$, $<f_{z_n}, z_n> \ge b^2 - \varepsilon$, $<f_{y_n}, z_m> \le \|f_{y_n}\| \cdot \|z_m\| \le ab + \varepsilon$, and $<f_{z_m}, y_n> \le \|f_{z_m}\| \cdot \|y_n\| \le ab + \varepsilon$. Thus,

$$<f_{y_n} - f_{z_m}, y_n - z_m> \ge a^2 + b^2 - 2ab - 4\varepsilon = 8\varepsilon - 4\varepsilon = 4\varepsilon ,$$ which

would contradict the assumption that $\lim_{n,m\to\infty} <f_{x_n} - f_{x_m}, x_n - x_m> = 0$.

We conclude that $\lim_{n\to\infty} \|x_n\|$ must therefore exist. □

<u>8.52 Definition.</u> Let E be a Banach space. The norm on E will be called locally uniformly convex if, for each $\varepsilon \in (0,2]$ and each x with $\|x\| = 1$, there exists $\delta = \delta(\varepsilon,x) > 0$ such that, for each $y \in E$ with $\|y\| = 1$ and $\|x+y\| > 2 - \delta$, we have $\|x-y\| < \varepsilon$.

<u>8.53 Lemma.</u> Assume E is reflexive. Then the norm on E is locally uniformly convex $\Leftrightarrow$ the dual norm on E' is C^1.

<u>Proof.</u> More standard Banach space functional analysis. □

Thus a Banach space E is variational $\Leftrightarrow$ it is reflexive and the norms on both E and E' are locally uniformly convex.

<u>8.54 Example.</u> Let $p \in (1,\infty)$, and let S be any measure space. Then $L^p(S,R)$ is uniformly convex, hence locally uniformly convex, and hence a variational space.

<u>8.55 Proposition.</u> Let E be a locally uniformly convex Banach space, and assume $\{x_n\}_{n\in N}$ is a sequence in E which converges weakly to x. Then $\lim_{n\to\infty} x_n = x \Leftrightarrow \lim_{n\to\infty} \|x_n\| = \|x\|$.

<u>Proof.</u> Obviously, if $\{x_n\}_{n\in N}$ converges to x in norm, then $\lim_{n\to\infty} \|x_n\| = \|x\|$. To see the converse, assume that $\lim_{n\to\infty} \|x_n\| = \|x\|$. If $\|x\| = 0$ we are done, so assume that $\|x\| \neq 0$. By changing a finite number of terms of the sequence if necessary, we may assume that $\|x_n\| \neq 0$ for each $n \in N$. Let $y = \frac{x}{\|x\|}$, and for each $n \in N$, let $y_n = \frac{x_n}{\|x_n\|}$; then $\|y\| = 1$, $\|y_n\| = 1$ for all $n \in N$, and by the weak continuity of scalar multiplication, $\{y_n\}_{n\in N}$ converges weakly to y. Let $f_y : E \to R$ be a linear functional such that $\|f_y\| = 1$ and $\langle f_y, y\rangle = 1$. Then $\langle f_y, y+y_n\rangle = \langle f_y, y\rangle + \langle f_y, y_n\rangle \to 2\langle f_y, y\rangle = 2$. This implies that $\|y + y_n\| \to 2$ and hence, since E is locally uniformly convex, that $\lim_{n\to\infty} y_n = y$. Finally, by continuity of scalar multiplication, $x = \|x\|\cdot y = \lim_{n\to\infty} (\|x_n\|\cdot y_n)$

$= \lim_{n\to\infty} x_n$. □

We have now finished our discussion of norms on Banach spaces and are ready to return to critical point theory. We will need to present one more lemma, and then we will be ready to state and prove the abstract version of our main result in the calculus of variations.

<u>8.56 Lemma</u>. Let M be a $\mathcal{U}^\infty$ manifold such that $J(M)$ has a Bw* Finsler structure, E a variational space, and $F : M \to E$ a $\mathcal{U}^1$ map. Let $f : M \to \mathbb{R}$ be the function given by $f(x) = \|F(x)\|^2$. Then:

(a) f is strongly C^1 .

(b) f is weakly coercive .

<u>Proof</u>. Since $\| \quad \|^2 : E \to \mathbb{R}$ is C^1 , and since $\mathcal{U}^1$ maps are strongly C^1 , $f = \| \quad \|^2 \circ F$ is strongly C^1 .

To see (b), assume that $\{x_n\}_{n\in N}$ is a weakly convergent sequence such that $\|Df(x_n)\| \to 0$, and let $x_0 = \lim_{n\to\infty} x_n$. Let X be a model for the component of M which contains x_0, and let (U, φ) be a chart on M which contains x_0 and such that $\varphi(U)$ is convex in X . By changing a finite number of terms in the sequence if necessary, we may assume that $x_n \in U$ for each $n \in N$. Let $y_n = \varphi(x_n)$, $y_0 = \varphi(x_0)$, $G = F \circ \varphi^{-1}$, and $g = f \circ \varphi^{-1}$. Also, let $J(\varphi(U))$ have the Finsler structure induced by $J(M)$ via pull-back by $J(\varphi^{-1})$. Then $g = \| \quad \|^2 \circ G$, $\|Df(x)\| = \|Dg(\varphi(x))\|$ for each $x \in U$, and we can finish the proof by showing that $\|Dg(y_0)\| = 0$ and that $G(y_0) = \lim_{n\to\infty} G(y_n)$.

Let $A = \{y_n\}_{n\geq 0}$, and let K be the closed convex hull of A in X . Then, since A is compact and a subset of $\varphi(U)$, we know that K is also compact and a subset of $\varphi(U)$. In addition, we know that the restriction to K of the Finsler structure on $\varphi(U)$ is equivalent to the Finsler structure which K inherits from any norm on X'' . So, at this point, let us replace the Finsler structure on K with this equivalent "linear" Finsler structure.

Let $h : E \to E'$ be the duality map for E. Then, for each $y \in \varphi(U)$, $w \in X$, we have $Dg_y(w) = D(\| \quad \|^2)_{G(y)} \circ DG_y(w)$
$= \langle h_{G(y)}, DG_y(w)\rangle$.

Now, for each $m,n \in N$,

$$G(y_m) - G(y_n) = DG_{y_n}(y_m - y_n) + \int_0^1 (DG_{ty_m + (1-t)y_n} - DG_{y_n})(y_m - y_n)\,dt .$$

Since G is of differentiability class $\mathcal{U}^1$, there exists a semi-norm λ on X such that, for each $u,v \in K$, $z \in X$,

$$\|(DG_u - DG_v)(z)\| \le \lambda(u - v)\cdot\|z\| + \|u - v\|\cdot\lambda(z) .$$

Thus $\|(DG_{ty_m + (1-t)y_n} - DG_{y_n})(y_m - y_n)\| \le \lambda(ty_m - ty_n)\cdot\|y_m - y_n\|$

$$+ \|ty_m - ty_n\|\cdot\lambda(y_m - y_n) = 2t\cdot\|y_m - y_n\|\lambda(y_m - y_n) .$$

Now we use the weak continuity of G: since G is weakly continuous, $\{G(y_n)\}_{n\in N}$ converges weakly to $G(y_o)$, and hence the sequence is bounded in E.

Now we come to the key inequality of this proof:

$$|\langle h_{G(y_m)} - h_{G(y_n)}, G(y_m) - G(y_n)\rangle|$$

$$\le |\langle h_{G(y_m)}, G(y_n) - G(y_m)\rangle| + |\langle h_{G(y_n)}, G(y_m) - G(y_n)\rangle|$$

$$\le |\langle h_{G(y_m)}, DG_{y_m}(y_n - y_m)\rangle| +$$

$$|\langle h_{G(y_m)}, \int_0^1 (DG_{ty_n + (1-t)y_m} - DG_{y_m})(y_m - y_n)\,dt\rangle|$$

$$+ |\langle h_{G(y_n)}, DG_{y_n}(y_m - y_n)\rangle| +$$

$$|\langle h_{G(y_n)}, \int_0^1 (DG_{ty_m + (1-t)y_n} - DG_{y_n})(y_m - y_n)\,dt\rangle|$$

$$\le |Dg_{y_m}(y_n - y_m)| + \|h_{G(y_m)}\|\cdot\|y_m - y_n\|\cdot\lambda(y_m - y_n)$$

$$+ |Dg_{y_n}(y_m - y_n)| + \|h_{G(y_n)}\|\cdot\|y_m - y_n\|\cdot\lambda(y_m - y_n) .$$

Since $\{y_n\}_{n\in N}$ converges weakly to y, $\{G(y_n)\}_{n\in N}$ converges weakly to $G(y)$, and hence $\{h_{G(y_n)}\}_{n\in N}$ is a bounded set of functionals on

E . Also, since λ is continuous on X , $\lim_{n,m\to\infty} \lambda(y_m - y_n) = 0$. Adding to this the fact that $\lim_{n\to\infty} \|Dg_{y_n}\| = 0$, we conclude that each of the four terms at the end of the above inequality converges to zero as $m,n\to\infty$. Thus, since E is a variational space, this implies that $\{G(y_n)\}_{n\in N}$ converges to $G(y_\circ)$ in the strong topology on E .

To see that $Dg(y_\circ) = 0$, let $z \in X$, and note that

$$Dg_{y_\circ}(z) = \langle h_{G(y_\circ)}, DG_{y_\circ}(z)\rangle = \lim_{n\to\infty} \langle h_{G(y_\circ)}, DG_{y_n}(z)\rangle$$

$$= \lim_{n\to\infty} (\langle h_{G(y_n)}, DG_{y_n}(z)\rangle + \langle h_{G(y_\circ)} - h_{G(y_n)}, DG_{y_n}(z)\rangle) .$$

Now, $\langle h_{G(y_n)}, DG_{y_n}(z)\rangle = Dg_{y_n}(z)$, and since $\|Dg_{y_n}\| \to 0$, this term goes to zero as $n \to \infty$. As for the second term, since $\{G(y_n)\}_{n\in N}$ converges strongly to $G(y_\circ)$, $\|h_{G(y_\circ)} - h_{G(y_n)}\| \to 0$. Thus $Dg_{y_\circ}(z) = 0$ for all $z \in X$, i.e. $Dg_{y_\circ} = 0$. □

<u>8.57 Theorem</u>. Let M be a complete $\mathcal{U}^\infty$ manifold, E a variational space, $F : M \to E$ a $\mathcal{U}^1$ map, and $p > 1$. Then the function $f : M \to R$ which is defined by $f(x) = \|F(x)\|^p$ has the following properties:

(a) f is bounded below by zero.

(b) f is lower semi-continuous.

(c) f is strongly C^1 .

(d) f is bounded on compact subsets of M .

(e) f satisfies Condition (WC) $\Leftrightarrow$ F is proper.

<u>Proof</u>. (a) is obvious. As for (b), we need to show that $f^{-1}((c,\infty))$ is open in M for each $c \in R$. Well, if $c < 0$, then $f^{-1}((c,\infty)) = M$. If $c > 0$, then $f^{-1}((c,\infty)) = F^{-1}(E - B_0(c^{1/p}))$. And if $c = 0$, $f^{-1}((0,\infty)) = F^{-1}(E - \{0\})$. Since F is weakly continuous, and since $B_0(r)$ is weakly closed in E for each $r \geq 0$, we are done.

Part (c) follows since $f = \| \quad \|^p \circ F$. F is strongly C^1 since it is $\mathcal{U}^1$, and $\| \quad \|^p$ is C^1 by 8.44(d) .

Let A be compact in M. Then $F(A)$ is compact in E, and hence bounded in E, and hence $\| \quad \|^p \circ F$ is bounded on A.

Finally, (e) will follow from Theorem 8.24 if we can show that f is weakly coercive. So let $\{x_n\}_{n \in N}$ be a weakly convergent sequence in M such that $\|Df_{x_n}\| \to 0$, and let $x_o = \lim_{n\to\infty} x_n$. If $\lim_{n\to\infty} f(x_n) = 0$, then by the lower semi-continuity of f, we conclude that $f(x_o) = 0$, and hence that $Df(x_o) = 0$, since in this case f has a minimum at x_o.

So now assume that $\limsup_{n \in N} f(x_n) = b > 0$. Since f is bounded on weakly compact sets, we automatically have $b < \infty$. Let $\{y_n\}_{n\in N}$ be a subsequence of $\{x_n\}_{n\in N}$ such that $\lim_{n\to\infty} f(y_n) = b$ and such that $f(y_n) > 0$ for all $n \in N$. Since f is lower semi-continuous, $f^{-1}(0)$ is closed in M, and hence $M - f^{-1}(0)$ is an open submanifold of M. Now, define $g : M \to R$ by $g(x) = \|F(x)\|^2$ for each $x \in M$. Then, for each $z \in M - f^{-1}(0)$, $g(z) = (f(z))^{2/p}$ and $Dg_z = \frac{2}{p}(f(z))^{\frac{2}{p}-1} Df_z$. Since f is bounded away from zero on $\{y_n\}_{n\in N}$, this implies that $\|Dg_{y_n}\| \to 0$. But we know from Lemma 8.56 that g is weakly coercive on M, so $g(z_o) = \lim_{n\to\infty} g(y_n) = b^{2/p}$, and $Dg(x_o) = 0$. Thus $x_o \in M - f^{-1}(0)$. But, for each $z \in M - f^{-1}(0)$, $f(z) = (g(z))^{p/2}$ and $df_z = \frac{p}{2}(g(z))^{\frac{p}{2}-1} Dg_z$, which implies that $f(x_o) = b$ and $Df_{x_o} = 0$. To see that $f(x_n) \to f(x_o)$, note that, since f is lower semi-continuous, $\limsup_{n\in N} f(x_n) = b = f(x_o) \leq \liminf_{n\in N} f(x_o)$, which implies that $\lim_{n\to\infty} f(x_n)$ exists and is equal to $f(x_o)$. □

<u>8.58 Corollary</u>. Let M be a $\mathcal{U}^\infty$ manifold, E a variational space, $i : M \to E$ a $\mathcal{U}^1$ embedding, and $p > 1$. Define $f : M \to R$ by $f(x) = \|i(x)\|^p$, and let $J(M)$ have the Finsler structure induced by the embedding. Then f satisfies Condition (C).

<u>Proof</u>. Since $i : M \to E$ is an embedding, $J(M)$ is complete in the induced Bw* Finsler structure. Thus Theorem 8.57 implies that f satisfies Condition (WC). Now, let $\{x_n\}_{n\in N}$ be a weakly

convergent sequence such that $\|Df_{x_n}\| \to 0$, and let $x_\circ = \lim_{n\to\infty} x_n$. Then, since f is weakly coercive, $\|i(x_n)\| \to \|i(x_\circ)\|$. And since i is weakly continuous, $\{i(x_n)\}_{n\in N}$ converges weakly to $i(x_\circ)$. Thus, by Proposition 8.55, $i(x_n)\}$ converges strongly to $i(x_\circ)$. Finally, since a $\mathcal{U}^1$ embedding is automatically a strong C^1 embedding, we conclude that $\{x_n\}_{n\in N}$ converges strongly to $x_\circ$, and hence that f satisfies Condition (C). □

Let $x \in M$, and let X be the model space for the component of M which contains x. If $F : M \to E$ is a $\mathcal{U}^1$ embedding, as in the above corollary, then Df(x) gives a weak embedding of X into E, and hence implies that X'' is a reflexive Banach space. In the more general case considered in Theorem 8.57, DF(x) need not be a linear embedding for each $x \in M$. However, if F is weakly proper, then it seems reasonable to conjecture that there must be at least one point x in each component of M at which DF(x) is an embedding. If this conjecture can be proved, it would imply that each component of J(M) is modeled on a reflexive Banach space, and hence that Theorem 8.57 cannot be applied to show the existence of critical points for nonlinear functionals on manifolds which are modeled on nonreflexive Banach function spaces.

The original result concerning the existence of critical points for a functional on a Banach manifold was obtained independently by Palais and Smale for the case where M is a Banach manifold modeled on Hilbert space, and $f : M \to \mathbb{R}$ is a C^2 function with nondegenerate critical points. Their result is a natural infinite-dimensional generalization of the classical result of M. Morse, and is an improvement over Corollary 8.38 because in general it gives more precise information about the number of critical points, and because "most" C^2 functionals can be shown to have nondegenerate critical points.

Now, unfortunately, Condition (WC) is not enough to show that a nonlinear functional satisfies the conclusions of Morse-Palais-Smale theory, even if the functional has nondegenerate critical points; we need to know that the functional satisfies Condition (C). However, for a functional of the sort described in Theorem 8.57, we are able to show that, if it has nondegenerate critical points, then it satisfies Condition (C).

Incidentally, the concept of nondegeneracy for a critical point has been generalized to Banach spaces (see, for example, [1], [25] and [34]). However, the concept is of limited interest unless it is possible to show that "most" functions, in some sense, have only nondegenerate critical points. To show this latter result, it is necessary to assume that the domain space E of the function is isomorphic to E'. Since the only reflexive function spaces used in PDE's which are isomorphic to their respective duals happen also to be Hilbert spaces, there seems to be little point in treating the more general case here.

<u>8.59 Lemma</u>. Let X_1, X_2 be bw* spaces, U open in X_1, $f : U \to X_2$ a $\mathcal{U}^2$ map, and C a compact subset of U. Then there is a semi-norm λ on X_1 such that, for each $x,y \in C$, $v,z \in X_1$,

$$\|D^2f_y(v,z) - D^2f_x(v,z)\| \le \|y-x\|\cdot\lambda(v)\cdot\|z\| + \lambda(y-x)\cdot\|v\|\cdot\|z\| .$$

<u>Proof</u>. Since f is $\mathcal{U}^2$, the map $Tf: U\times X_1 \to X_2\times X_2$ given by $Tf(x,z) = (f(x), Df_x(z))$ is $\mathcal{U}^1$. Thus $p_2 \circ Tf$ is $\mathcal{U}^1$, where $p_2 : X_2\times X_2 \to X_2$ is projection onto the second factor.

Let K be the unit ball in X_1. Then $C \times K$ is compact in $U \times X_1$. Hence there is a semi-norm λ on X_1 such that, if $x,y \in C$, $z,u \in K$, $v,w \in X_1$:

$$\|D(p_2 \circ Tf)_{(y,z)}(v,w) - D(p_2 \circ Tf)_{(x,u)}(v,w)\|$$
$$\le (\lambda(y-x) + \lambda(z-u))\cdot(\|v\| + \|w\|) + (\|y-x\| + \|z-u\|)\cdot(\lambda(v) + \lambda(w)) .$$

Now, $D(p_2 \circ Tf)_{(y,z)}(v,w) - D(p_2 \circ Tf)_{(x,u)}(v,w)$

$$= D^2f_y(v,z) + Df_y(w) - D^2f_x(v,u) - Df_x(w) .$$

Evaluating the inequality at $w = 0$, and letting $u = z$, we have

$$\|D^2f_y(v,z) - D^2f_x(v,z)\| \le \lambda(y-x)\cdot\|v\| + \|y-x\|\cdot\lambda(v) .$$

Now, if z is an arbitrary nonzero vector in X_1, then

$\frac{z}{\|z\|} \in K$, and hence

$$\|D^2f_y(v,\frac{z}{\|z\|}) - D^2f_x(v,\frac{z}{\|z\|})\| \le \lambda(y-x)\cdot\|v\| + \|y-x\|\cdot\lambda(v) .$$

Since D^2f_y and D^2f_x are bilinear, the desired inequality now follows for $z \neq 0$ by multiplying both sides of this inequality by $\|z\|$. And if $z = 0$, the proof of the inequality is obvious. □

8.60 Lemma. Let X_1, X_2 be bw* spaces, U open in X_1, $F : U \to X_2$ a $\mathcal{U}^1$ map, $x_0 \in U$, and $\{x_n\}_{n\in N}$ a sequence in U which converges to x_0. Assume that $\{F(x_n)\}_{n\in N}$ converges strongly to $F(x_0)$, and assume in addition that X_2'' is reflexive. Then, for each $\varepsilon > 0$, there exists $n_0 \in N$ such that, for all $n > n_0$ and $t \in [0,1]$,

$$\|F(tx_n + (1-t)x_0) - F(x_0)\| < \varepsilon .$$

Proof. Since X_2'' is reflexive we may assume that it has a locally uniformly convex norm, and we may assume that U is convex. Let $\varepsilon > 0$, and assume for the moment that the conclusion of the theorem does not hold for this ε. Then there exists a subsequence $\{y_n\}_{n\in N}$ of the given sequence and a sequence $\{t_n\}_{n\in N}$ of real numbers in the unit interval such that, if we let $z_n = t_n\cdot y_n + (1-t_n)x_0$, then $\|F(z_n) - F(x_0)\| \ge \varepsilon$ for all $n \in N$. Note that $\{z_n\}_{n\in N}$ converges weakly to x_0.

Let $A = \{x_n\}_{n\geq 0}$, and let K be the closed convex hull of A. Then K is a compact subset of U, so there exists a seminorm λ on X_2 such that, for each $x,y \in K$, $z \in X_1$,

$$\|(DF_y - DF_x)(z)\| \le \lambda(y-x)\cdot\|z\| + \|y-x\|\cdot\lambda(z) .$$

$$\text{Now,}\quad F(z_n) - F(x_0) = DF_{x_0}(z_n - x_0) + \int_0^1 (DF_{tz_n+(1-t)x_0} - DF_{x_0})(z_n - x_0)\,dt$$

Let $\gamma_n = \frac{\|z_n - x_0\|}{\|y_n - x_0\|}$, let $s = \gamma_n t$, and note that

$$tz_n + (1-t)x_0 = x_0 + t(z_n - x_0) = x_0 + \frac{s}{\gamma_n}(z_n - x_0) = x_0 + s(y_n - x_0) =$$

$sy_n + (1-s)x_o$. Applying the Change of Variables Theorem,

$$\int_0^1 (DF_{tz_n+(1-t)x_o} - DF_{x_o})(z_n - x_o)\,dt =$$

$$\int_0^{\gamma_n} (DF_{sy_n+(1-s)x_o} - DF_{x_o})(z_n - x_o)\left(\frac{1}{\gamma_n}\right)ds$$

$$= \int_0^{\gamma_n} (DF_{sy_n+(1-s)x_o} - DF_{x_o})\ (y_n - x_o)\,ds\ .$$

Thus, we have:

$$\left\|\int_0^1 (DF_{tz_n+(1-t)x_o} - DF_{x_o})(z_n - x_o)\,dt\right\|$$

$$\le \int_0^1 \|DF_{sy_n+(1-s)x_o} - DF_{x_o})(y_n - x_o)\|ds \le \int_0^1 2s\cdot\|y_n - x_o\|\cdot\lambda(y_n - x_o)\,ds$$

$$= \|y_n - x_o\|\cdot\lambda(y_n - x_o)\ .$$

Therefore, $\|F(z_n)\| \le \|F(x_o) + DF_{x_o}(z_n - x_o)\| + \|y_n - x_o\|\cdot\lambda(y_n - x_o)$.

Similarly, $\|F(y_n)\| \ge \|F(x_o) + DF_{x_o}(y_n - x_o)\| - \|y_n - x_o\|\cdot\lambda(y_n - x_o)$.

Now, $F(x_o) + DF_{x_o}(z_n - x_o)$ is on the straight line segment between $F(x_o)$ and $F(x_o) + DF_{x_o}(y_n - x_o)$. Since the norm is a convex function on X_2'' ,

$$\|F(x_o) + DF_{x_o}(z_n - x_o)\| \le \max\{\|F(x_o)\|\ ,\ \|F(x_o) + DF_{x_o}(y_n - x_o)\|\}$$

Thus, $\|F(z_n)\| \le 2\lambda(y_n - x_o)\cdot\|y_n - x_o\| + \max\{\|F(x_o)\|\ ,\ \|F(y_n)\|\}$.

Since $\lim_{n\to\infty} \lambda(y_n - x_o) = 0$ and $\lim_{n\to\infty} \|F(y_n)\| = \|F(x_o)\|$, $\limsup_{n\in N} \|F(z_n)\| \le \|F(x_o)\|$. But $\{F(z_n)\}_{n\in N}$ converges weakly to $F(x_o)$, and since the norm function is lower semi-continuous in the weak topology, $\|F(x_o)\| \le \liminf_{n\in N} \|F(z_n)\|$. Thus $\lim_{n\to\infty} \|F(z_n)\| = \|F(x_o)\|$; and since the norm on X_2'' is locally uniformly convex, Proposition 8.55 implies that $\{F(z_n)\}_{n\in N}$ converges strongly to $F(x_o)$. But this contradicts the assumption involved in our definition of $\{z_n\}_{n\in N}$. Therefore no such sequence can exist, and we are done. □

8.61 Lemma. Let M be a $\mathcal{C}^\infty$ manifold modeled on a separable Hilbert space, and assume that $J(M)$ has a Bw* Riemannian metric.

Let H be a Hilbert space, $F : M \to H$ a $\mathcal{U}^2$ map, and $f : M \to \mathbb{R}$ the map defined by $f(x) = \|F(x)\|^2$. Then f is strongly C^2. Furthermore, if the critical points of f are nondegenerate, then f is coercive.

Proof. Since F is $\mathcal{U}^2$, it is automatically strongly C^2; and $\| \quad \|^2 : H \to \mathbb{R}$ is C^∞. So f is the composition of two strongly C^2 maps, and hence is strongly C^2.

Now, assume that the critical points of f are nondegenerate. Let $\{x_n\}_{n\in N}$ be a weakly convergent sequence such that $\|Df_{x_n}\| \to 0$, and let $x_0 = \lim_{n\to\infty} x_n$. Choose a chart (U,φ) such that $x_0 \in U$ and such that $\varphi(U)$ is convex. Let X be the model space for M, so $\varphi(U)$ is open in X. By changing a finite number of terms of the sequence if necessary, we may assume that $x_n \in U$ for each $n \in N$. Letting $G = F \circ \varphi^{-1}$, $g = f \circ \varphi^{-1}$, $y_n = \varphi(x_n)$ for each $n \geq 0$, and letting $J(\varphi(U))$ have the Riemannian metric induced from the metric on $J(M)$ via φ, we have $\|Dg_{\varphi(x)}\| = \|Df_x\|$ for each $x \in U$, so $\|Dg_{y_n}\| \to 0$.

Let $A = \{y_n\}_{n\geq 0}$, and let K be the closed convex hull of A. Then K is compact and a subset of $\varphi(U)$, and hence the Riemannian metric on K is equivalent to the Riemannian metric which K inherits as a subset of X''. Thus, we will now replace the Riemannian metric on K with this equivalent metric. Finally, since $Df_{x_0} = 0$, $D^2(g)_{y_0} = D(D(f\circ\varphi^{-1}))_{y_0} = D(((Df)\circ\varphi^{-1})\cdot D\varphi^{-1})_{y_0}$

$$= ((D^2f)\circ\varphi^{-1})_{y_0}\cdot(D\varphi^{-1},D\varphi^{-1})_{y_0} + ((Df)\circ\varphi^{-1})_{y_0}\cdot(D^2\varphi^{-1})_{y_0}$$

$$= (D^2f)_{x_0}\cdot(D\varphi^{-1},D\varphi^{-1})_{y_0},$$ which shows that y is a nondegenerate critical point for g.

For each $y \in K$, $z \in X$, $Dg_y(z) = 2\langle G(y), DG_y(z)\rangle$, so $D^2g_y(z,w) = 2\langle DG_y(z), DG_y(w)\rangle + 2\langle G(y), D^2G_y(z,w)\rangle$.

Thus, we have:

$$(*) \quad D^2g_y(z,w) - D^2g_{y_0}(z,w) = 2\langle DG_y(z), DG_y(w)-DG_{y_0}(w)\rangle$$
$$+ 2\langle DG_y(z) - DG_{y_0}(z), DG_{y_0}(w)\rangle$$
$$+ 2\langle G(y), D^2G_y(w,z)-D^2G_{y_0}(w,z)\rangle + 2\langle G(y) - G(y_0), D^2G_{y_0}(w,z)\rangle .$$

Now, consider $D^2g_{y_0}$: since this bilinear form is nondegenerate, there exists an isomorphism $A : X \to X$ such that $D^2g_{y_0}(z,w) = \langle Az,w\rangle$ for each $z,w \in X$ (in fact, since the bilinear form is symmetric, so is the operator). For each $n \in N$, let $v_n = A^{-1}(y_n - y_0)$. Then $D^2g_{y_0}(y_n-y_0, v_n) = \langle y_n-y_0, Av_n\rangle = \langle y_n-y_0, y_n-y_0\rangle = \|y_n-y_0\|^2$, and $\{v_n\}_{n\in N}$ converges weakly to zero.

Let $\varepsilon > 0$. Since DG is $\mathcal{U}$-regulated, there exists $n_1 \in N$ such that, for all $n \geq n_1$ and for all $t \in [0,1]$,

$$\|DG_{ty_n+(1-t)y_0}(y_n-y_0) - DG_{y_0}(y_n-y_0)\| < \varepsilon \quad \text{and}$$

$$\|DG_{ty_n+(1-t)y_0}(v_n) - DG_{y_0}(v_n)\| < \varepsilon .$$

By Lemma 8.59, we can find $n_2 \in N$ such that, for all $n \geq n_2$ and $t \in [0,1]$, $\|D^2G_{ty_n+(1-t)y_0}(y_n-y_0, v_n) - D^2G_{y_0}(y_n-y_0, v_n)\| < \varepsilon$. By Lemma 8.60, we can find $n_3 \in N$ such that, for all $n \geq n_3$ and $t \in [0,1]$, $\|G(ty_n+(1-t)y_0) - G(y_0)\| < \varepsilon$.

Since $Dg_{y_0} = 0$, we have the following:

$$Dg_{y_n}(v_n) = Dg_{y_n}(v_n) - Dg_{y_0}(v_n) = \int_0^1 D^2g_{ty_n+(1-t)y_0}(y_n-y_0, v_n)\,dt$$

$$= D^2g_{y_0}(y_n-y_0, v_n)$$

$$+ \int_0^1 (D^2g_{ty_n+(1-t)y_0}(y_n-y_0, v_n) - D^2g_{y_0}(y_n-y_0, v_n))\,dt$$

The previous paragraph implies that the integral goes to zero as $n \to \infty$. Thus, there exists $n_0 \in N$ such that, for all $n \geq n_0$,

$Dg_{y_n}(v_n) \geq \frac{1}{2} D^2g_{y_0}(y_n-y_0, v_n) = \frac{1}{2}\|y_n-y_0\|^2$. But since $\{v_n\}_{n\in N}$ is bounded in X, and since $\|Dg_{y_n}\| \to 0$, we conclude that $\|y_n-y_0\| \to 0$. Thus f is coercive. □

<u>8.62 Theorem</u>. Let M be a $\mathcal{U}^\infty$ manifold modeled on a separable Hilbert space, and assume that $J(M)$ has a complete Bw* Riemannian metric. Let H be a Hilbert space, $F : M \to H$ a $\mathcal{U}^2$ map, and $f : M \to R$ the map defined by $f(x) = \|F(x)\|^2$. Assume in addition that the critical points of f are nondegenerate. Then f satisfies Condition (C) $\Leftrightarrow$ F is proper.

Proof. Since f is bounded on compact subsets of M, and since f is coercive by the above lemma, this theorem follows from Theorem 8.16. □

The methods which enable us to go from the theorems of this chapter to actual estimates of the number of critical points which a nonlinear functional must have, are standard differential topology of Banach manifolds and algebraic topology, and will not be repeated here. We will, however, quote the results and give references for the interested reader.

8.63 Definition. Let S be an arcwise connected topological space. Then cuplong (S) is defined to be the supremum of all integers n such that, for some field F and elements $\alpha_i \in H^{k_i}(S,F)$ with $k_i > 0$, $1 \le i \le n$, $\alpha_i \cup \dots \cup \alpha_n \neq 0$.

8.64 Theorem. Let S be arcwise connected. Then $\mathrm{cat}(S) \geq 1 + \mathrm{cuplong}(S)$.

Proof. See [30]. □

Let M be a connected Bw* manifold. By Theorem 7.11, M and $J(M)$ have the same homotopy type, and hence the same homology and cohomology, so it doesn't matter whether we use the cohomology of M or the cohomology of $J(M)$ in the computation of cuplong $(J(M))$. In the general case, when M is not connected, we can use Theorem 8.64 to estimate the category of each component of M, and simply add the results together to get an estimate for $\mathrm{cat}(M)$.

The value of Theorem 8.62 is that the number of critical points of a C^2 function f on a Hilbert manifold M with a complete Riemannian metric can be estimated via the Morse inequalities if the function is bounded below, has only nondegenerate critical points, and satisfies Condition (C). Without going into details (see [29] or [30] for details), the basic result is that the number of critical points of f is at least $\Sigma_{n=0}^{\infty}\beta_n(M)$, where $\beta_n(M)$ is the nth Betti number of M. To see how this compares with

Corollary 8.38, note that we can define $\mathbb{R}$-cuplong(M) in the same way as the definition of cuplong(M) , but restricting ourselves to $H(M,\mathbb{R})$, i.e. to cohomology with real coefficients. Then it is straightforward that $\mathbb{R}$-cuplong(M) $\le \Sigma_{n=0}^{\infty}\beta_n(M)$. Now, this does not really show that cat(M) $\le \Sigma_{n=0}^{\infty}\beta_n(M)$. But it does provide an indication that, for most applications, when it comes to estimating cat(M) , the Morse inequalities (if applicable) provide a better estimate of the number of critical points of f then does Lusternik-Schnirelman theory.

9. AN APPLICATION TO THE CALCULUS OF VARIATIONS

This chapter will consist of results about function spaces and differential operators which are needed in nonlinear variational theory, and will conclude with applications of Theorems 8.57 and 8.62 to a specific class of nonlinear variational problems.

We will begin by applying Theorems 8.57 and 8.62 to obtain some basic results in the calculus of variations. These applications will be immediate once we have made a series of observations about function spaces and nonlinear differential operators. This material was developed in large part by Palais, and can be found in Chapters 15-17 and 19 of his Foundations of Global Nonlinear Analysis. Thus we will content ourselves here with a restatement of the relevant definitions and results, with proofs omitted in selected instances where no modification of Palais' treatment is required. The reader who is dissatisfied with the treatment presented here is advised to consult the above-cited chapters of Palais' monograph.

We return to the notation of Chapters 1 and 6 . Let M be a compact C^{∞} manifold, $\mathcal{M}$ a LTS section functor on $FB(M)$, E a fiber bundle over M , and $f \in \mathcal{M}(E)$.

9.1 Definition. We define a subset $\mathcal{M}_{\partial f}(E)$ of $\mathcal{M}(E)$ by letting $\mathcal{M}_{\partial f}(E)$ be the closure in $\mathcal{M}(E)$ of the set of $g \in \mathcal{M}(E)$ for which there exists a neighborhood $U_g \subseteq M$ of ∂M such that $f|_{U_g} = g|_{U_g}$.

We will now show, via a series of lemmas, that $\mathcal{M}_{\partial f}(E)$ is a submanifold of $\mathcal{M}(E)$.

9.2 Notation. Let ξ be a vector bundle over M . Then $\mathcal{M}_0(\xi)$ will be used to denote $\mathcal{M}_{\partial 0}(\xi)$, where 0 denotes the zero section of ξ . Note that $\mathcal{M}_0(\xi)$ is a closed linear subspace of $\mathcal{M}(\xi)$.

9.3 Lemma. Let ξ be a vector bundle over M, $f, g \in \mathcal{M}(\xi)$. Then add : $\mathcal{M}_{\partial f}(\xi) \oplus \mathcal{M}_{\partial g}(\xi) \to \mathcal{M}_{\partial(f+g)}(\xi)$.

Proof. Note that addition is a continuous map from $\mathcal{M}(\xi) \oplus \mathcal{M}(\xi)$ to $\mathcal{M}(\xi)$. Let $f', g' \in \mathcal{M}(\xi)$, such that f and f' agree on a neighborhood of ∂M, and g and g' agree on a neighborhood of ∂M: then $\text{add}(f' \oplus g') = f' + g' \in \mathcal{M}_{\partial(f+g)}(\xi)$. Since $\text{add}^{-1}(\mathcal{M}_{\partial(f+g)}(\xi))$ is closed in $\mathcal{M}(\xi) \oplus \mathcal{M}(\xi)$, the conclusion follows. □

9.4 Lemma. Let ξ be a vector bundle over M, and let $f \in \mathcal{M}(\xi)$. Then $\mathcal{M}_{\partial f}(\xi) = f + \mathcal{M}_{\circ}(\xi)$.

Proof. $f + \mathcal{M}_0(\xi) \subset \mathcal{M}_{\partial f}(\xi)$ by the preceding lemma. Similarly, since $\text{add}(\mathcal{M}_{\partial f}(\xi) \oplus \mathcal{M}_{\partial(-f)}(\xi)) \subset \mathcal{M}_{\circ}(\xi)$, $\mathcal{M}_{\partial f}(\xi) - f \subset \mathcal{M}_{\circ}(\xi)$, which implies that $\mathcal{M}_{\partial f}(\xi) \subset f + \mathcal{M}_{\circ}(\xi)$. □

9.5 Lemma. Let ξ be a vector bundle over M, $f, g \in \mathcal{M}(\xi)$, and assume that $g \in \mathcal{M}_{\partial f}(\xi)$. Then $\mathcal{M}_{\partial g}(\xi) = \mathcal{M}_{\partial f}(\xi)$.

Proof. By the preceding lemma, there exists $h \in \mathcal{M}_{\circ}(\xi)$ for which $g = f + h$. Thus $\mathcal{M}_{\partial g}(\xi) = g + \mathcal{M}_{\circ}(\xi) = (f + h) + \mathcal{M}_{\circ}(\xi)$ $= f + (h + \mathcal{M}_{\circ}(\xi)) = f + \mathcal{M}_{\circ}(\xi) = \mathcal{M}_{\partial f}(\xi)$. □

Let $C^{\infty}_{\circ}(\xi) = \{s \in C^{\infty}(\xi) \mid (\text{support } s) \cap \partial M = \emptyset\}$.

9.6 Lemma. Let ξ be a vector bundle over M. If $C^{\infty}(\xi)$ is a dense subset of $\mathcal{M}(\xi)$, then $C^{\infty}_{\circ}(\xi)$ is a dense subset of $\mathcal{M}_{\circ}(\xi)$.

Proof. Let f be a section of $\mathcal{M}(\xi)$ for which (support f) $\cap\ \partial M = \emptyset$. Choose a C^{∞} real-valued function α on M such that $\alpha(\text{support } f) = 1$ and $(\text{support } \alpha) \cap \partial M = \emptyset$, and let $m_{\alpha} : \mathcal{M}(\xi) \to \mathcal{M}(\xi)$ be multiplication by α. Then $m_{\alpha}(C^{\infty}(\xi)) \subset C^{\infty}_{\circ}(\xi)$. Thus, since $C^{\infty}(\xi)$ is dense in $\mathcal{M}(\xi)$, $m_{\alpha}(\mathcal{M}(\xi)) = m_{\alpha}(\overline{C^{\infty}(\xi)}) \subset \overline{C^{\infty}_{\circ}(\xi)}$. But since $m_{\alpha}(f) = f$, we have $f \in \overline{C^{\infty}_{\circ}(\xi)}$. Since this is true for each $f \in \mathcal{M}(\xi)$ for which (support f) $\cap\ \partial M = \emptyset$, the closure of these functions must also be a subset of $\overline{C^{\infty}_{\circ}(\xi)}$, i.e. $\mathcal{M}_{\circ}(\xi) \subset \overline{C^{\infty}_{\circ}(\xi)}$. And since $C^{\infty}(\xi) \subset \mathcal{M}(\xi)$, $\overline{C^{\infty}_{\circ}(\xi)} \subset \mathcal{M}_{\circ}(\xi)$. □

9.7 Remark. Let E be a fiber bundle over M, $f,g \in \mathcal{M}(E)$. If there is a neighborhood $U \subseteq M$ of ∂M such that $f|_U = g|_U$, then $\mathcal{M}_{\partial f}(E) = \mathcal{M}_{\partial g}(E)$.

9.8 Remark. Let Z be a topological space, W open in Z, and A a subset of Z. Then the closure of $A \cap W$, regarded as a subspace of W, is equal to $\bar{A} \cap W$, where $\bar{A}$ is the closure of A in Z.

9.9 Lemma. Let E be a fiber bundle over M, ξ an open vector subbundle of E, and $f \in \mathcal{M}(\xi)$. Then $\mathcal{M}_{\partial f}(\xi) = \mathcal{M}_{\partial f}(E) \cap \mathcal{M}(\xi)$.

Proof. Immediate from previous remark. □

9.10 Lemma. Let E be a fiber bundle over M, ξ an open vector subbundle of E, and $f \in \mathcal{M}(E)$ such that $f(\partial M) \subset \xi$. Then there exists $g \in \mathcal{M}(\xi)$ and an open neighborhood $U \subseteq M$ of ∂M for which $f|_U = g|_U$.

Proof. Since f is continuous, there is an open neighborhood V of ∂M such that $f(V) \subset \xi$. Choose an open neighborhood U of ∂M such that $\bar{U} \subset V$. Let $A = f(\bar{U})$, and let $C = \pi^{-1}(M - V)$, where $\pi : \xi \to M$ is the bundle projection. Since A is compact, C is closed, and $A \cap C = \emptyset$, there exists a C^∞ real-valued function $\beta : \xi \to [0,1]$ such that $\beta(A) = 1$ and $\beta(C) = 0$. Furthermore, since A is compact and ξ is locally compact, we may assume that support(β) is compact. Define a fiber-preserving map $\alpha : \xi \to \xi$ by $\alpha(v) = (\beta(v))\cdot v$ for each $v \in \xi$. Then α is C^∞. Furthermore, since $\alpha(v) = 0_{\pi(v)}$ for all vectors except those contained in support(β), and since support(β) is compact, we may extend α to a C^∞ fiber-preserving map from E to ξ, which we will also denote by α, by defining $\alpha(x) = 0_{p(x)}$ for each $x \in E - \xi$, where $p : E \to M$ is the bundle projection. Thus we have a map $\mathcal{M}(\alpha)$ from $\mathcal{M}(E)$ to $\mathcal{M}(\xi)$. If we let $g = \mathcal{M}(\alpha)(f)$, then $g \in \mathcal{M}(\xi)$, and for each $x \in U$, $g(x) = \alpha \circ f(x) = \alpha(f(x)) = \beta(x)\cdot f(x) = f(x)$. □

9.11 Proposition. Let E be a fiber bundle over M, and $f \in \mathcal{M}(E)$. Assume that $g \in \mathcal{M}_{\partial f}(E)$. Then $\mathcal{M}_{\partial g}(E) = \mathcal{M}_{\partial f}(E)$.

Proof. We first show $\mathcal{M}_{\partial g}(E) \subset \mathcal{M}_{\partial f}(E)$. To see this, let $h \in \mathcal{M}_{\partial g}(E)$, and choose an open vector subbundle ξ of E with $h \in \mathcal{M}(\xi)$. Since $h \in \mathcal{M}_{\partial g}(E)$, $h|_{\partial M} = g|_{\partial M}$. Similarly, since $g \in \mathcal{M}_{\partial f}(E)$, $g|_{\partial M} = f|_{\partial M}$. Thus $g(\partial M) \subset \xi$ and $f(\partial M) \subset \xi$. So Lemma 9.10 implies the existence of $g', f' \in \mathcal{M}(\xi)$ and a neighborhood U of ∂M such that $g'|_U = g|_U$ and $f'|_U = f|_U$. Note that $\mathcal{M}_{\partial g'}(E) = \mathcal{M}_{\partial g}(E)$ and $\mathcal{M}_{\partial f'}(E) = \mathcal{M}_{\partial f}(E)$, so $g' \in \mathcal{M}_{\partial f'}(\xi)$. By Lemma 9.5, $h \in \mathcal{M}_{\partial f'}(\xi) \subset \mathcal{M}_{\partial f'}(E)$, which implies $\mathcal{M}_{\partial g}(E) \subset \mathcal{M}_{\partial f}(E)$.

The above part of the proof will also imply that $\mathcal{M}_{\partial f}(E) \subset \mathcal{M}_{\partial g}(E)$ if we can show that $f \in \mathcal{M}_{\partial g}(E)$. To see this, let η be an open vector subbundle of E with $f \in \mathcal{M}(\eta)$. Since $g(\partial M) = f(\partial M) \subset \eta$, another application of the above lemma yields the existence of $g' \in \mathcal{M}(\eta)$ and a neighborhood V of ∂M such that $g'|_V = g|_V$. This implies $g' \in \mathcal{M}_{\partial g}(E) \subset \mathcal{M}_{\partial f}(E)$, hence $g' \in \mathcal{M}_{\partial f}(\eta)$ by Lemma 9.9. Hence Lemma 9.5 implies $f \in \mathcal{M}_{\partial g'}(\eta) \subset \mathcal{M}_{\partial g'}(E) = \mathcal{M}_{\partial g}(E)$. □

9.12 Lemma. Let E be a fiber bundle over M, and $f \in \mathcal{M}(E)$. Assume that $g \in \mathcal{M}_{\partial f}(E)$, and that ξ is an open vector subbundle of E for which $g \in \mathcal{M}(\xi)$. Then $\mathcal{M}_{\partial f}(E) \cap \mathcal{M}(\xi) = \mathcal{M}_{\partial g}(\xi) = g + \mathcal{M}_0(\xi)$.

Proof. By Proposition 9.11, $\mathcal{M}_{\partial f}(E) = \mathcal{M}_{\partial g}(E)$. But by Lemma 9.9, $\mathcal{M}_{\partial g}(E) \cap \mathcal{M}(\xi) = \mathcal{M}_{\partial g}(\xi) = g + \mathcal{M}_0(\xi)$. □

9.13 Theorem. Let $\mathcal{M}$ be a D section functor on FB(M), E a fiber bundle over M, and $f \in \mathcal{M}(E)$. Then $\mathcal{M}_{\partial f}(E)$ is a closed C^∞ submanifold of $\mathcal{M}(E)$.

Proof. Let ξ, η be an open vector subbundles such that $\mathcal{M}_{\partial f}(E) \cap \mathcal{M}(\xi) \cap \mathcal{M}(\eta) \neq \emptyset$. Choose $g \in \mathcal{M}_{\partial f}(E) \cap \mathcal{M}(\xi) \cap \mathcal{M}(\eta)$, let $T_g : \mathcal{M}(\xi) \to \mathcal{M}(\xi)$ be affine translation by g, and let $T'_{(-g)} : \mathcal{M}(\eta) \to \mathcal{M}(\eta)$ be affine translation by $(-g)$. Then

$T'_{(-g)} \circ \mathcal{M}(\mathrm{id}) \circ T_g$ maps an open subset of $\mathcal{M}_o(\xi)$ bijectively to an open subset of $\mathcal{M}_o(\eta)$. Note that T_g and $T'_{(-g)}$ are both C^∞ diffeomorphisms. And $\mathcal{M}(\mathrm{id}) : \mathcal{M}(\xi) \to \mathcal{M}(\eta)$ is a C^∞ diffeomorphism. Since $\mathcal{M}_o(\xi)$ is a closed subspace of $\mathcal{M}(\xi)$, it is a D space, and hence the composition $T'_{(-g)} \circ T_g$ is a C^∞ diffeomorphism of an open subset of $\mathcal{M}_o(\xi)$ with an open subset of $\mathcal{M}_o(\eta)$. □

<u>9.14 Corollary</u>. Let ω be a $\mathcal{U}$ section functor on FB(M) , E a fiber bundle over M , and $f \in \omega(E)$. Then $\omega_{\partial f}(E)$ is a closed $\mathcal{U}^\infty$ submanifold of $\omega(E)$.

<u>Proof</u>. It suffices to observe that the coordinate change in $\omega_{\partial f}(E)$ which appears in the proof of the above theorem is a $\mathcal{U}^\infty$ diffeomorphism, and this follows since $\omega(\mathrm{id})$ is $\mathcal{U}^\infty$ and the translations T_g and $T'_{(-g)}$ are $\mathcal{U}^\infty$. □

We turn now to differential operators and jet bundles. The notation will be the same as that employed in the second half of Chapter 6.

<u>9.15 Remark</u>. Let M be a compact manifold, E a fiber bundle over M , ξ an open vector subbundle of E , and $k \in N$. Then $J^k(\xi)$ is an open vector subbundle of E .

Let $\mathcal{M}$ be a section functor on FB(M) . The above remark, which is a consequence of the way the bundle $J^k(E)$ is defined, gives us a covering of $\mathcal{M}(J^k(E))$ by a set of charts derived from charts on $\mathcal{M}(E)$.

<u>9.16 Lemma</u>. Let ω be a $\mathcal{U}$ section functor on FB(E) , $k \in N$. Then the k-jet extension map $j_k : \omega_k(E) \to \omega(J^k(E))$, introduced in the definition of derivative section functors, is a $\mathcal{U}^\infty$ map.

<u>Proof</u>. Note that ω_k is a $\mathcal{U}$ section functor by Theorem 6.37. It suffices to let ξ be any open vector subbundle of E , and to show that the map $j_k : \omega_k(\xi) \to \omega(J^k(\xi))$ is $\mathcal{U}^\infty$. But this follows because j_k is a linear map from $\omega_k(\xi)$ to $\omega(J^k(\xi))$. □

9.17 Definition. Let M be a compact manifold, E_1 and E_2 objects of $FB(M)$. A function $D : C^\infty(E_1) \to C^\infty(E_2)$ will be called a smooth differential operator of order k from E_1 to E_2 if it can be factored as

$$C^\infty(E_1) \xrightarrow{j_k} C^\infty(J^k(E_1)) \xrightarrow{C^\infty(f)} C^\infty(E_2)$$

where $f : J^k(E_1) \to E_2$ is a smooth bundle morphism over M.

9.18 Theorem. Let $D : C^\infty(E_1) \to C^\infty(E_2)$ be a smooth k^{th} order differential operator such that $D = C^\infty(f) \circ j_k$, where $f : J^k(E_1) \to E_2$ is a smooth bundle morphism. Then D extends to a $\mathcal{U}^\infty$ map from $\omega_k(E_1)$ to $\omega(E_2)$ given by $D(s) = f(j_k(s))$ for each $s \in \omega_k(E_1)$.

Proof. We observed in Lemma 9.16 that $j_k : \omega_k(E_1) \to \omega(J^k(E_1))$ is $\mathcal{U}^\infty$. And since ω is a $\mathcal{U}$ section functor, $\omega(f) : \omega(J^k(E_1)) \to \omega(E_2)$ is $\mathcal{U}^\infty$. Thus the map $\omega(f) \circ j_k$ is $\mathcal{U}^\infty$. □

Corollary. Let M be a compact n-dimensional manifold, E_1 and E_2 objects of $FB(M)$, $p \in (1,\infty)$, $k \in N$, and $r > \frac{n}{p}$. Assume that $D : C^\infty(E_1) \to C^\infty(E_2)$ is a smooth k^{th} order differential operator. Then D extends to a unique $\mathcal{U}^\infty$ map from $L^p_{k+r}(E_1)$ to $L^p_r(E_2)$.

Proof. We need only verify the uniqueness of the extension, which follows because $C^\infty(E_1)$ is dense in $L^p_{k+r}(E_1)$. □

The above corollary is of limited interest in the calculus of variations. In the calculus of variations we are presented with a kth order differential operator $D : C^\infty(E) \to C^\infty(\eta)$, where η is a vector bundle over M, and where D extends to a map from $L^p_k(E)$ to $L^p_0(\eta)$. Now, it is well-known and easy to see that the general nonlinear operator does not admit such an extension. So our first task is to introduce an abstract class of operators which admit such extensions, and which includes many of the examples of operators which appear in applications. Such a class is presented in Chapter 16 of

Palais' Foundations of Global Non-Linear Analysis. We will briefly describe it here.

Let n, p, k, ℓ, w be integers, with $0 \le \ell < k \le w$, and $n, p > 0$. We will use the letter β to denote an n-multi-index $(b_1,\ldots,b_n)$ with $b_i \geq 0$ for each $0 \le i \le n$, and we will define B to be the set of all ordered w-tuples of n-multi-indices $(\beta_1,\ldots,\beta_w)$ such that $|\beta_i| > \ell$ or $|\beta_i| = 0$ for each i, and such that $|\beta_1| +\ldots+ |\beta_w| \le w$. If β is the multi-index $(0,\ldots,0)$, and $f : R^n \to R$ is a C^∞ function, then by $\frac{\partial^{|\beta|} f}{\partial x^\beta}$ we will mean (for this definition only) the constant function 1. And finally, let A denote the set of functions from $\{1,\ldots,w\}$ to $\{1,\ldots,p\}$. With these conventions, we can make the following definition:

9.19 Definition. Let M be a compact n-dimensional submanifold of R^n, $D : C^\infty(M,R^p) \to C^\infty(M,R)$ a smooth k^{th} order differential operator. We will say that D is a polynomial differential operator of weight w with respect to derivatives of order $> \ell$ if, for each $\lambda \in B$, $\alpha \in A$, there is an ℓth order operator $F_{\lambda\alpha}: C^\infty(M,R^p) \to C^\infty(M,R)$ such that

$$Ds = \sum_{\substack{\lambda\in B\\ \alpha\in A}} F_{\lambda\alpha}(s) \cdot \frac{\partial^{|\beta_1|} s_{\alpha(1)}}{\partial x^{\beta_1}} \cdot \ \ldots \ \cdot \frac{\partial^{|\beta_w|} s_{\alpha(w)}}{\partial x^{\beta_w}}$$

for each $s \in C^\infty(M,R^p)$, where s_j denotes the jth coordinate function of the vector-valued function s.

A kth order operator $D : C^\infty(M,R^p) \to C^\infty(M,R^q)$ will be called a polynomial operator from R^p to R^q of weight w with respect to derivatives of order $> \ell$ if $Ds = (D_1 s,\ldots,D_q s)$ for each $s \in C^\infty(M,R^p)$, where each $D_i: C^\infty(M,R^p) \to C^\infty(M,R)$ is a polynomial operator of weight w with respect to derivatives of order $> \ell$. We will abbreviate this symbolically by $D \in OP_k^{w,\ell}(M\times R^p, M\times R^q)$.

Note that the operators $\{F_{\lambda\alpha} | \lambda \in B, \alpha \in A\}$, if they exist, are not unique, since for simplicity of description we have allowed the indexing sets A and B to be much larger than necessary.

9.20 Definition. Let M be a compact contractible n-dimensional submanifold of R^n, E a p-dimensional fiber bundle over M, η a q-dimensional vector bundle over M, and $D : C^\infty(E) \to C^\infty(\eta)$ a kth order differential operator. Let ℓ, w be integers with $0 \leq \ell < k \leq w$. We will say that D is a kth order polynomial operator of weight w with respect to derivatives of order $> \ell$ (abbreviated symbolically by $D \in OP_k^{w,\ell}(E,\eta)$) if, for each open vector subbundle ξ of E and trivializations $\varphi : \xi \approx M \times R^p$, $\psi : \eta \approx M \times R^q$, the induced k^{th} order operator from $C^\infty(M,R^p)$ to $C^\infty(M,R^q)$ is in $OP_k^{w,\ell}(M\times R^p, M\times R^q)$.

The assumption in the above definition that M is contractible is made merely to ensure that vector bundles over M are trivializable, which simplifies the statement of the definition.

9.21 Theorem. Let M be a compact contractible n-dimensional submanifold of R^n, E a fiber bundle over M, η a vector bundle over M, and D is smooth k^{th} order differential operator from $C^\infty(E)$ to $C^\infty(\eta)$. In order that $D \in OP_k^{w,\ell}(E,\eta)$, it is sufficient that, for each $s \in C^\infty(E)$, there is an open vector bundle neighborhood ξ of s and trivializations $\varphi : \xi \approx M\times R^p$, $\psi : \eta \approx M\times R^q$, such that the induced operator from $C^\infty(M,R^p)$ to $C^\infty(M,R^q)$ is in $OP_k^{w,\ell}(M\times R^p, M\times R^q)$.

Proof. See [33]. □

9.22 Lemma. Let M be a compact contractible n-dimensional submanifold of R^n, $D \in OP_k^{w,\ell}(M\times R^p, M\times R^q)$, and $\varphi : M \to R^n$ a diffeomorphic embedding. Then the induced operator F from $C^\infty(\varphi(M),R^p)$ to $C^\infty(\varphi(M),R^q)$, given by $F(s) = D(s\circ\varphi)\circ\varphi^{-1}$, is in $OP_k^{w,\ell}(\varphi(M)\times R^p, \varphi(M)\times R^q)$.

9.23 Definition. Let M be a compact n-dimensional manifold, E a fiber bundle over M, η a vector bundle over M, and $D : C^\infty(E) \to C^\infty(\eta)$ a kth order operator. We will say that $D \in OP_k^{w,\ell}(E,\eta)$ if, for each chart (U,φ) on M and compact contractible n-dimensional submanifold N of U, the induced operator from $C^\infty((\varphi^{-1})^*(E|_N))$ to $C^\infty((\varphi^{-1})^*(\eta|_N))$ is of weight w with respect to derivatives of order $> \ell$.

9.24 Proposition. Let M be a compact n-dimensional manifold, E a fiber bundle over M, η a vector bundle over M, and $D : C^\infty(E) \to C^\infty(\eta)$ a kth order operator. In order that $D \in OP_k^{w,\ell}(E,\eta)$, it is sufficient that, for each $x \in M$, there is a chart (U,φ) and a compact contractible n-dimensional submanifold N of U such that $x \in N$ and such that the induced operator from $C^\infty((\varphi^{-1})^*(E|_N))$ to $C^\infty((\varphi^{-1})^*(\eta|_N))$ is of weight w with respect to derivatives of order $> \ell$.

Proof. Immediate from Lemma 9.22 and Theorem 9.21. □

The above proposition implies that there are many differential operators in $OP_k^{w,\ell}(E,\eta)$, since it implies that such operators may be constructed locally in vector bundle neighborhoods of E and patched together by a partition of unity on E to yield a globally-defined operator. In addition it implies that, if E is a subbundle of F and $D : C^\infty(F) \to C^\infty(\eta)$ is in $OP_k^{w,\ell}(F,\eta)$, then $D|_{C^\infty(E)}$ is in $OP_k^{w,\ell}(E,\eta)$.

9.25 Theorem. Let $p,q \in [1,\infty)$, $0 \le \ell < k - \frac{n}{p} < r \le k$, and $r \le m$. Assume $D \in OP_r^{w,\ell}(E,\eta)$, where E and η are bundles over a compact n-dimensional manifold M. Then D extends to a strongly C^∞ map from $L_k^p(E)$ to $L_{k-m}^q(\eta)$, provided

$$\frac{n}{p} \ge (k-m) + \frac{w}{r}\left(\frac{n}{p} - (k-r)\right).$$

Proof. See Theorem 16.10 of [33]. □

9.26 Corollary. Let $p \in [1,\infty)$, $0 \le \ell < k - \frac{n}{p}$, and assume $D \in OP_k^{k,\ell}(E,\eta)$. Then D extends to a strongly C^∞ map from $L_k^p(E)$ to $L_0^p(\eta)$.

9.27 Corollary. Let $p \in [1,\infty)$, $0 \le \ell < \frac{k-1}{k}(k-\frac{n}{p})$, and assume $D \in OP_{k-1}^{k,\ell}(E,\eta)$. Then D extends to a strongly C^∞ map from $L_{k-1}^{\frac{pk}{k-1}}(E)$ to $L_0^p(\eta)$.

Our next major goal is to prove an improved version of Corollary 9.26. The reader should note that Corollary 9.26 is concerned with polynomial operators of weight and order k . This class of operators seems to be the smallest easily-defined class of operators which includes the linear differential operators of order k between vector bundles and which is closed under nonlinear coordinate changes in the domain bundle. Corollary 9.26 is due to Palais. Corollary 9.27, which will be a major tool in the proof of our strengthened version of 9.26, is due to K. Uhlenbeck. In fact, the proof of the main result of this chapter (Theorem 9.34) was motivated by the argument Uhlenbeck employed in Theorem 12.1 of [45].

<u>9.28 Lemma</u>. Let X_1, X_2, X_3, X_4 be bw* spaces, $h : X_2 \times X_3 \to X_4$ a continuous bilinear map. Let $E_i = X_i''$, for each i , and assume that there is a continuous linear embedding $i : X_1 \to E_2$. Then the induced bilinear map $\mathbf{h} : X_1 \times X_3 \to X_4$ is $\mathcal{U}^\infty$.

<u>Proof</u>. By Proposition 6.7, it suffices to show that h is $\mathcal{U}^1$ on $X_1 \times X_3$. Since $h : X_2 \times X_3 \to X_4$ is continuous, it follows that the induced bilinear map $h : E_2 \times E_3 \to E_4$ sends bounded sets to bounded sets, and hence is continuous. By renorming E_4 if necessary, we may assume that $\|h(x,y)\|_4 \le 1$ whenever $\|x\|_2 \le 1$ and $\|y\|_3 \le 1$. Choose a semi-norm λ on X_1 such that $\|z\|_2 \le \lambda(z)$ for all $z \in X_1$. Then, if $(u_i, v_i) \in X_1 \times X_3$,

$$\|(Dh_{(u_1,v_1)} - Dh_{(u_2,v_2)})(u_3,v_3)\|_4$$

$$= \|Dh_{(u_1,v_1)}(u_3,v_3) - Dh_{(u_2,v_2)}(u_3,v_3)\|_4$$

$$= \|h(u_1,v_3) + h(u_3,v_1) - h(u_2,v_3) - h(u_3,v_2)\|_4$$

$$= \|h(u_1-u_2,v_3) + h(u_3,v_1,-v_2)\|_4 \le \|h(u_1-u_2,v_3)\|_4 + \|h(u_3,v_1-v_2)\|_4$$

$$\le \lambda(u_1-u_2)\cdot\|v_3\|_3 + \|v_1-v_2\|_3\cdot\lambda(u_3)$$, which implies that h is $\mathcal{U}^1$ on $X_1 \times X_3$. □

<u>9.29 Lemma</u>. Let $p \in (1,\infty)$, $k > \frac{n}{p}$. Then multiplication extends to a $\mathcal{U}^\infty$ map from $L^p_k(D^n,R) \times L^p_0(D^n,R)$ to $L^p_0(D^n,R)$.

<u>Proof</u>. Let $j \in (\frac{n}{p},k)$, define q by $\frac{1}{q} = 1 - \frac{1}{p}$, and let $\delta \in (0,\frac{n}{q})$. Then, by part (3) of Theorem 9.5 of [33] , multiplication extends to a continuous bilinear map from $L^p_j(D^n,R) \times L^p_\delta(D^n,R)$ to $L^q_\delta(D^n,R)$. Taking the dual action of $L^p_j(D^n,R)$, and noting that $(L^q_\delta(D^n,R))' = L^p_{-\delta}(D^n,R)$, we have that multiplication extends to a bilinear map from $L^p_j(D^n,R) \times L^p_{-\delta}(D^n,R)$ to $L^p_{-\delta}(D^n,R)$.

Let $i \in (j,k)$. Then multiplication extends to a strongly continuous bilinear map from $L^p_i(D^n,R) \times L^p_0(D^n,R)$ to $L^p_0(D^n,R)$, and hence bounded sets get mapped to bounded sets. Thus we have the following commutative diagram, where the horizontal maps are continuous multiplications, the vertical maps are completely continuous inclusions, and multiplication maps bounded sets to bounded sets:

$$\begin{array}{ccc}
L^p_i(D^n,R) \times L^p_0(D^n,R) & \xrightarrow{\ m\ } & L^p_0(D^n,R) \\
\downarrow & & \downarrow \\
L^p_j(D^n,R) \times L^p_{-\delta}(D^n,R) & \xrightarrow{\ m\ } & L^p_{-\delta}(D^n,R)
\end{array}$$

It follows immediately by the same argument as was used in Chapter 1 in the construction of Bw* section functors that $m : L^p_i(D^n,R) \times L^p_0(D^n,R) \to L^p_0(D^n,R)$ is weakly continuous. Finally, since the inclusion of $L^p_k(D^n,R)$ into $L^p_i(D^n,R)$ is completely continuous, the desired conclusion follows from Lemma 9.28 . □

<u>9.30 Lemma</u>. Let X_1, X_2, X_3 be bw* spaces, $E_i = E_i''$ for each i , U open in X_1 , and V open in E_2 . Assume that $g : V \to E_3$ is a strongly C^∞ map, and that $i : X_1 \to E_2$ is a continuous linear inclusion. Then $g \circ i : U \to X_3$ is $\mathcal{U}^\infty$.

<u>Proof</u>. Note first that $g \circ i : U \to X_3$ is weakly C^∞ . Let D be a compact set in U , and $n \in N$. Then, since D is compact in V , $Dg(i(D))$ is bounded in $L(E_2,E_3)$. Thus there exists $c > 0$ such that, for each $x \in D$, $v_1,\ldots,v_n \in X_1$, $\|D^n g_x(v_1,\ldots,v_n)\|_3 \le c\|v_1\|_2 \cdots \|v_n\|_2$. Since the E_2-norm is continuous on X_1 , Proposition 6.10 implies that $g \circ i$ is $\mathcal{U}^\infty$. □

9.31 Theorem. Let $0 \le \ell < \frac{k-1}{k}(k - \frac{n}{p})$, and let $D \in OP_k^{k,\ell}(D^n \times R^q, D^n \times R)$. Then D extends to a $\mathcal{U}^\infty$ map from $L_k^p(D^n, R^q)$ to $L_0^p(D^n, R)$.

Proof. We may write $D = A + \sum_{\substack{|\beta|=k \\ 1\le j\le q}} D_{\beta j}$, where A is an operator of weight k and order $k-1$ mod ℓ-derivatives, and each D_β is a monomial operator of the form $D_{\beta j}(s) = F_{\beta j}(s)\cdot\frac{\partial^{|\beta|} s_j}{\partial x^\beta}$ with $F_{\beta j}$ an operator of order ℓ. Now, each $F_{\beta j}$ induces a $\mathcal{U}^\infty$ map from $L_k^p(D^n, R^q)$ to $L_{k-\ell}^p(D^n, R)$ by Theorem 6.40, and $\frac{\partial^{|\beta|}}{\partial x^\beta}$ induces a $\mathcal{U}^\infty$ map from $L_k^p(D^n, R)$ to $L_0^p(D^n, R)$ since it is linear. Since Lemma 9.29 says that multiplication is a $\mathcal{U}^\infty$ map from $L_\ell^p(D^n, R) \times L_0^p(D^n, R)$ to $L_0^p(D^n, R)$, and since $\mathcal{U}^\infty$ maps are closed under composition, we conclude that each $D_{\beta j}$ induces a $\mathcal{U}^\infty$ map from $L_k^p(D^n, R^q)$ to $L_0^p(D^n, R)$. As for the operator A, it induces a strongly C^∞ map from $L_{k-1}^{\frac{pk}{k-1}}(D^n, R^q)$ to $L_0^p(D^n, R)$ by Corollary 9.27. Since the inclusion $L_k^p(D^n, R^q) \to L_{k-1}^{\frac{pk}{k-1}}(D^n, R^q)$ is completely continuous (this is one of the basic Sobolev embedding theorems), an application of Lemma 9.30 concludes the proof of the theorem. □

9.32 Remark. If $\ell = 0$ and $k = 1$ in the above theorem, then we may write $D = A + \sum_{\substack{1\le i\le n \\ 1\le j\le q}} D_{ij}$, where A is an operator of order 0 and each D_{ij} is of the form $D_{ij}(s) = F_{ij}(s)\,\frac{\partial s_j}{\partial x_i}$. The conclusion of the theorem in this case is the same.

9.33 Theorem. Let M be a compact n-dimensional manifold, E a fiber bundle over M, η a vector bundle over M, $p \in (1,\infty)$, $k > \frac{n}{p}$, $\ell \ge 0$, and $D \in OP_k^{k,\ell}(E, \eta)$. If $\ell > 0$, assume in addition that $\ell < \frac{k-1}{k}(k - \frac{n}{p})$. Then D extends to a unique $\mathcal{U}^\infty$ map from $L_k^p(E)$ to $L_0^p(\eta)$.

Proof. Immediate from 9.31 and 9.32 . □

9.34 Theorem. Let M be a compact n-dimensional manifold, E a fiber bundle over M, η a vector bundle over M, and $s \in L_k^p(E)$. Let $p \in (1,\infty)$, $k > \frac{n}{p}$, $\ell \geq 0$, and $D \in OP_k^{k,\ell}(E,\eta)$. If $\ell > 0$, assume in addition that $\ell < \frac{k-1}{k}(k - \frac{n}{p})$. Also, assume that $L_{k,\partial s}^p(E)$ has a complete Bw* Finsler structure, and define $f : L_{k,\partial s}^p(E) \to \mathbb{R}$ by $f(\sigma) = \int_M \|D\sigma(x)\|^p dx$ for each $\sigma \in L_{k,\partial s}^p(E)$. Then:

(1) f is strongly C^1, weakly lower semi-continuous, and bounded below by zero.

(2) f satisfies Condition (WC) $\Leftrightarrow D : L_{k,\partial s}^p(E) \to L_0^p(\eta)$ is proper.

Proof. Immediate by applying Theorem 8.57 and Corollary 9.14 to Theorem 9.33. □

9.35 Corollary. Assume in the above theorem that $p = 2$, and that the Finsler structure on $L_{k,\partial s}^2(E)$ comes from a Bw* Riemannian metric. Then f is strongly C^∞. If, in addition, the critical points of f are nondegenerate, then f satisfies Condition (C) $\Leftrightarrow D : L_{k,\partial s}^2(E) \to L_0^2(\eta)$ is proper.

Proof. Immediate from Theorem 8.62. □

This concludes our discussion of variational calculus. Examples of energy functionals to which these theorems apply, as well as theorems concerning the regularity of the critical points of f, may be found in Chapters 13 and 14 of [45]. Also, the reader is advised to consult Chapter 19 of [33] for an earlier, non-intrinsic treatment of this material which motivated the treatments in both [45] and this monograph.

Louisiana State University

BIBLIOGRAPHY

[1] D. D. Ang and V. T. Tuan, "An elementary proof of the Morse-Palais lemma for Banach spaces", Proc. Amer. Math. Soc. 39 (1973), 642-644.

[2] C. Bessaga, "Every infinite-dimensional Hilbert space is diffeomorphic with its unit sphere", Bull. Acad. Polon. Sci. Math. Astron. Phys. 14 (1966), 27-31.

[3] T. S. Bolis, "Smooth partitions of unity in some infinite-dimensional manifolds", preprint.

[4] R. Bonic and J. Frampton, "Smooth functions on Banach manifolds", J. Math. Mech. 15 (1966), 877-898.

[5] R. Bonic, J. Frampton and A. Tromba, "Λ-manifolds", J. Functional Analysis 3 (1969), 310-320.

[6] J. Dieudonné, "Sur le théoreme de Lebesgue-Nikodym V", Canadian J. Math. 3 (1951), 129-139.

[7] J. Dowling, "Finsler geometry on Sobolev manifolds", Proc. Sympos. Pure Math., vol. XV, Amer. Math. Soc., Providence, R. I., 1970, 1-10.

[8] N. Dunford and J. T. Schwartz, Linear Operators, Part I, Interscience Publishers, New York, 1957.

[9] R. Edwards, Functional Analysis, Holt, Rinehart and Winston, New York, 1965.

[10] J. Eells, Jr., "On the geometry of function spaces", Symp. Inter. de Topologia Alg., Mexico, 1956, 303-308.

[11] J. Eells, Jr., "A setting for global analysis", Bull. Amer. Math. Soc. 72 (1966), 751-807.

[12] J. Frampton and A. Tromba, "On the classification of spaces of Holder continuous functions", preprint.

[13] R. Graff, "Elements of local non-linear functional analysis", Ph.D. Dissertation, Princeton University, Princeton, N. J., 1972.

[14] R. Graff, "On paracompact subsets of linear topological spaces", Proc. Amer. Math. Soc. 53 (1975), 361-366.

[15] A. Grothendieck, "Sur les espaces (F) et (DF)", Summa Brasiliensis 3 (1954), 57-123.

[16] J. Horvath, Topological Vector Spaces and Distributions, Addison-Wesley, Reading, Mass., 1966.

[17] H. Jacobowitz, "Implicit function theorems and isometric embeddings", Ann. of Math. 95 (1972), 191-225.

[18] J. Kelley, General Topology, Van Nostrand, Princeton, N. J., 1955.

[19] J. Kelley and I. Namioka, Linear Topological Spaces, Van Nostrand, Princeton, N. J., 1963.

[20] J. Kijowski and W. Szczyrba, "On differentiability in an important class of locally convex spaces", Studia Mathematica 30 (1968), 247-257.

[21] G. Kothe, Topological Vector Spaces, Springer-Verlag, Berlin, 1969.

[22] N. Krikorian, "Differentiable structures on function spaces", Trans. Amer. Math. Soc. 171 (1972), 67-82.

[23] N. Kuiper and Tersptra-Keppler, "Differentiable closed embeddings of Banach manifolds", Symposium in honour of Prof. G. de Rham, Springer-Verlag, 1970, 118-125.

[24] N. Kuiper, Variétés Hilbertiennes: Aspects Géometriques, Les Presses de l'Université de Montréal, Montreal, 1971.

[25] H. H. Kuo, "The Morse-Palais lemma on Banach spaces", Bull. Amer. Math. Soc. 80 (1974), 363-365.

[26] S. Lang, Introduction to Differentiable Manifolds, Interscience Publishers, New York, 1962.

[27] J. Milnor, Morse Theory, Princeton Univ. Press, Princeton, N. J., 1963.

[28] E. Nelson, Topics in Dynamics I: Flows, Princeton University Press, Princeton, N. J., 1969.

[29] R. S. Palais, "Morse theory on Hilbert manifolds", Topology 2 (1963), 299-340.

[30] R. S. Palais, "Lectures on the differentiable topology of infinite-dimensional manifolds", Mimeographed notes, Brandeis University, Waltham, Mass., 1964-1965.

[31] R. S. Palais, "Homotopy theory of infinite-dimensional manifolds", Topology 5 (1966), 1-16.

[32] R. S. Palais, "Lusternik-Schnirelman theory on Banach manifolds", Topology 5 (1966), 115-132.

[33] R. S. Palais, Foundations of Global Non-Linear Analysis, Benjamin, New York, 1968.

[34] R. S. Palais, "The Morse lemma for Banach spaces", Bull. Amer. Math. Soc. 75 (1969), 968-971.

[35] R. S. Palais, "Critical point theory and the minimax principle", Proc. Sympos. Pure Math., vol. XV, Amer. Math. Soc., Providence, R. I., 1970, 185-212.

[36] J. W. Robbin, "On the existence theorem for ordinary differential equations", Proc. Amer. Math. Soc. 19 (1968), 1005-1006.

[37] L. Rubel and J. Ryff, "The bounded weak-star topology and the bounded analytic functions", J. Functional Analysis 5 (1970), 167-183.

[38] J. T. Schwartz, "Generalizing the Lusternik-Schnirelman theory of critical points", Comm. Pure and Appl. Math. 17 (1964), 307-315.

[39] H. H. Schaefer, Topological Vector Spaces, Springer-Verlag, New York, N. Y., 1971.

[40] I. Singer, Bases in Banach Spaces I, Springer-Verlag, Berlin, 1970.

[41] S. Smale, "Morse theory and a non-linear generalization of the Dirichlet problem", Ann. of Math. 80 (1964), 382-396.

[42] N. E. Steenrod, "A convenient category of topological spaces", Michigan Math. J. 14 (1967), 133-152.

[43] W. Szczyrba, "Differentiation in locally convex spaces", Studia Mathematica 39 (1971), 289-306.

[44] S. L. Troyanski, "On locally uniformly convex and differentiable norms in certain non-separable Banach spaces", Studia Mathematica 37 (1971), 173-180.

[45] K. Uhlenbeck, "The calculus of variations and global analysis", Ph.D. Dissertation, Brandeis University, Waltham, Mass., 1968.

[46] J. N. Vladimirskii, "A criterion for the normability of a locally convex space", Mat. Issled. 6 (1971), vyp. 2 (20), 138-141, 163.

[47] K. Yosida, Functional Analysis, Springer-Verlag, New York, 1968.

[redacted]